LIMNOLOGICA

Ecology of Inland Waters

Vol. 19, No. 1

Biomanipulation

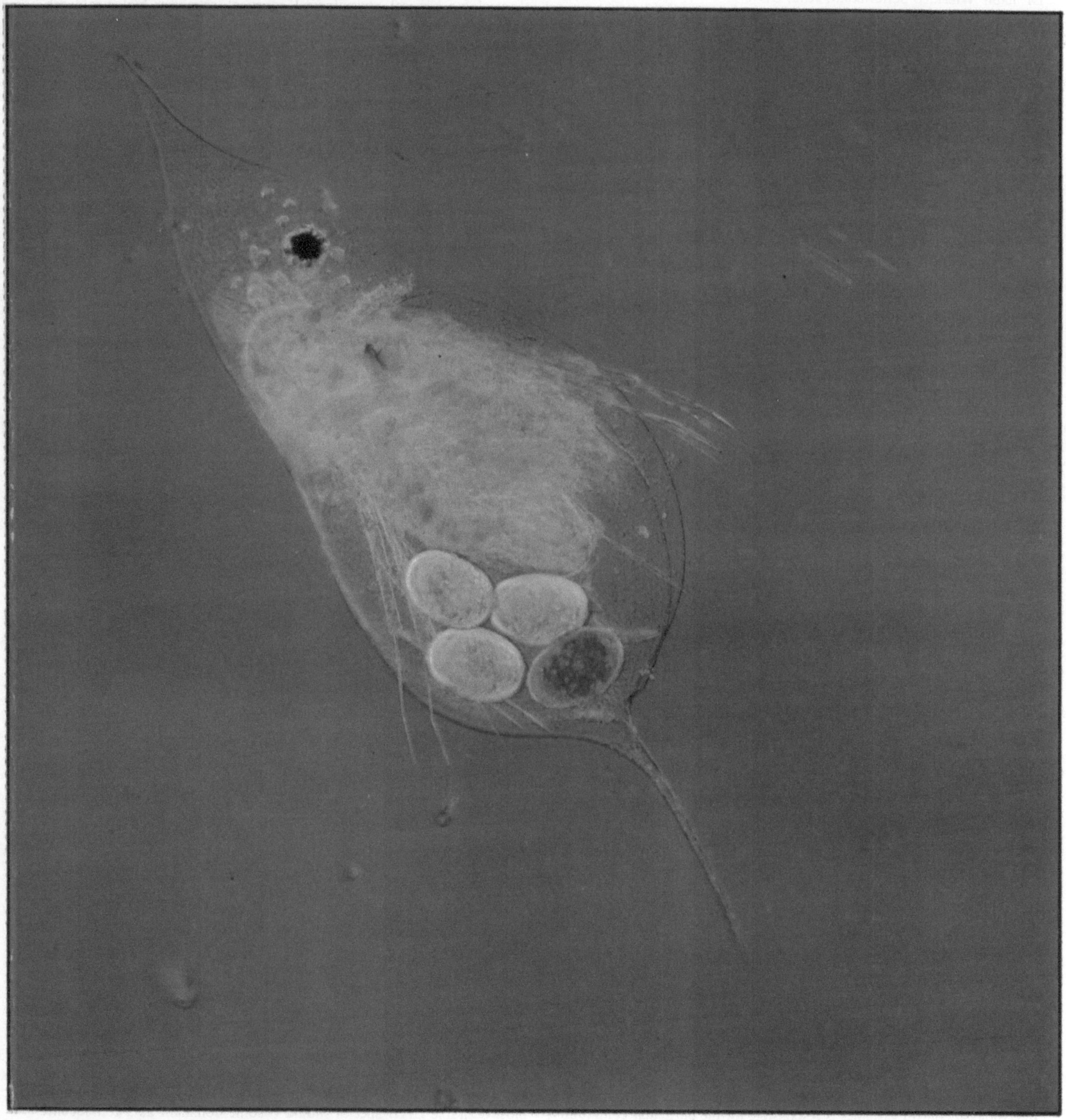

Akademie-Verlag Berlin

LIMNOLOGICA

Ecology of Inland Waters

Akademie der Wissenschaften der DDR
Forschungsbereich Biowissenschaften und Medizin
Zentralinstitut fur Mikrobiologie und Experimentelle Therapie Jena
Redaktion LIMNOLOGICA
DDR-6900 Jena, Beutenbergstraße 11

GENERAL INFORMATION

Scope LIMNOLOGICA were founded by Theodor Schrader in 1962 They will publish papers which fall in the scope of limnology, i e papers dealing with any aspects of ecology and hydrobiology of inland waters and adjacent biotopes, as well as reviews, discussions, and essays that develop limnological-ecological theory (including all qualitative and quantitative aspects of planktonic, periphytic, benthonic, neustonic, and nektonic organisms, nutrient and energy flow, bioindication, biomonitoring, water intern control mechanisms, ecological modelling and simulation, biological-ecotechnological water quality management). Proceedings of specialized symposia may also be accepted for publication, if the scientific standard set by the journal is met. Not acceptable is purely hydrological, hydrophysical, biochemical, physiological, genetical, or marine biological work, that, while using inland water test organisms, is unrelated to the scope of LIMNOLOGICA

Submission of articles. Manuscripts should be submitted to the editor-in-chief, the members of the editorial board, and the secretary

Publication schedule LIMNOLOGICA is published in one volume of nearly 480 pages per year Each volume comprises two issues

Subscriptions All issues are available upon your usual supplier all over the world If you can not get LIMNOLOGICA on this way, please directly send subscription orders to:
Akademie-Verlag Berlin
Leipziger Straße 3—4
DDR-1086 Berlin.

Readers in the FRG and Berlin(West), respectively, please send orders to
Kunst und Wissen
Erich Bieber
Wilhelmstr. 4—6
D-7000 Stuttgart 1.

Readers in other non-socialist countries please send orders to:
Buchexport
Volkseigener Außenhandelbetrieb der DDR
Leninstraße 16
DDR-7010 Leipzig

Cover Photo:

One of the main objectives of biomanipulation is to achieve the dominance of *Daphnia* in the zooplankton community Although the whole spectrum of ecological control variables is responsible for the particular community structure, the most important measure to create intentionally the dominance of *Daphnia* consists in controlling fish predation It is still unknown to a large extent which combination of controlling factors causes the dominance of a particular *Daphnia*-species But this knowledge is needed if not only small- and medium-sized species (such as *D. cucullata*, photo) but also the most desirable larger species should become dominant as a consequence of biomanipulation (Photo W Scheffler, Text: J. Benndorf)

Vorwort

Die LIMNOLOGICA wurden 1962 durch THEODOR SCHRÄDER begründet unter der Verantwortlichkeit der seinerzeitigen „Forschungsgemeinschaft der naturwissenschaftlichen, technischen und medizinischen Institute der Deutschen Akademie der Wissenschaften zu Berlin". Anfänglich mit Vorrang die Forschungsergebnisse der Mitarbeiter der Forschungsstelle für Limnologie publizierend, entwickelten sich die LIMNOLOGICA in den vergangenen 25 Jahren zu einem international anerkannten Publikationsorgan auf dem Gesamtgebiete der wissenschaftlichen Erforschung der Binnengewässer und der ihnen benachbarten Lebensräume. 1976 übernahm S. JOST CASPER die Verantwortung für die Herausgabe.
Die gegenwärtige Entwicklung der Limnologie und der ihr benachbarten Wissenschaftsdisziplinen hat einen solchen Standard erreicht, daß die ursprüngliche Konzeption nicht mehr aufrecht erhalten werden kann. Herausgeber und Verlag haben sich daher entschlossen, die Schriftenreihe zu aktualisieren und noch wirksamer als bisher in die internationalen Bemühungen um die Erforschung, den Schutz und die Sanierung der Binnengewässer und der ihnen benachbarten Lebensräume einzubeziehen. Sie hoffen und erwarten, daß die LIMNOLOGICA in Zukunft ihre Aufgaben noch besser bewältigen und vor allem nachhaltiges Interesse finden werden.

Herausgeber und Verlag

Preface

LIMNOLOGICA were founded by THEODOR SCHRÄDER in 1962 under the responsibility of the then „Forschungsgemeinschaft der naturwissenschaftlichen, technischen und medizinischen Institute der Deutschen Akademie der Wissenschaften zu Berlin". At first publishing with priority the results of the investigations made by the members of the staff of the then "Forschungsstelle für Limnologie" LIMNOLOGICA evolved into an international appreciated journal in the scientific field of fresh waters and their adjacent biocenoses. In 1976 S. JOST CASPER was appointed to the editorship.

The present high level of Limnology and its adjacent disciplins is not adequately reflected by the former conception of LIMNOLOGICA. Editors and publisher have made up their mind to bring up the journal to the modern topics of Limnology such as nutrient and energy flow, water intern control mechanisms, ecological modelling and simulation, and biological-ecotechnological water quality management including all qualitative and quantitative aspects of aquatic organisms. They hope all friends of LIMNOLOGICA will support their new editorial aims.

Editors and publisher

Biomanipulation

Editor: Jürgen Benndorf

XXIVth Congress of the International Association of theoretical and applied Limnology

---

August 13—19, 1989
Munich,
FRG

XXIV. Kongress der Internationalen Vereinigung für theoretische und angewandte Limnologie

---

13.—19. August 1989
München,
BRD

Steering Committee

Chairman:	Overbeck
Vice-Chairmen:	Fittkau
	Friedrich
	Siebeck
Treasurer:	Keiz
Members:	Elster
	Geller
	Hamm
	Kausch
	Immboden
	Lampert
	Müller, H.
	Nusch
	Ohle
	Ripl
	Saint-Paul
	Soeder
	Scharf, B.
	Schmitz
	Schwoerbel
	Tilzer
	Witzel
	Zwick

Limnologica (Berlin) 19 (1), 5–8, Juli 1988

"... the biocenotic relations between the fish stock and the plankton ... are at least as important as the influence of the physical and chemical factors for the formation of the plankton association ..."
HRBAČEK, DVOŘAKOVA, KOŘINEK und PROCHAZKOVA (1961)

Objectives and Unsolved Problems in Ecotechnology and Biomanipulation: A Preface

With one Figure.

Generally, there are two different strategies to control water quality in fresh waters:

(1) The strategy of reducing the external load of nutrients, toxic substances, organic matter or acid precipitation.
(2) The strategy of controlling internal ecological processes (ecotechnology).

With respect to toxic substances, organic wastes and acid precipitation, strategy (1) alone will provide acceptable solutions of the problems in a long-term scale. But, as far as eutrophication of lakes and reservoirs is concerned, a combination of both strategies leads to a better water quality and, consequently, to a lower cost/benefit ratio in the management of the water resource compared with the application of strategy (1) alone (Fig. 1). One of the most impressive case studies showing the success of this cooperation of external load reduction and internal ecological mechanisms is Lake Washington (EDMONDSON & LITT 1982) although the internal mechanisms were not used intentionally in this lake (EDMONDSON & ABELLA 1988). Furthermore, there are water bodies to which strategy (1) at present cannot be employed for economic, political or technological reasons. In these cases, strategy (2) offers a possibility to minimize the

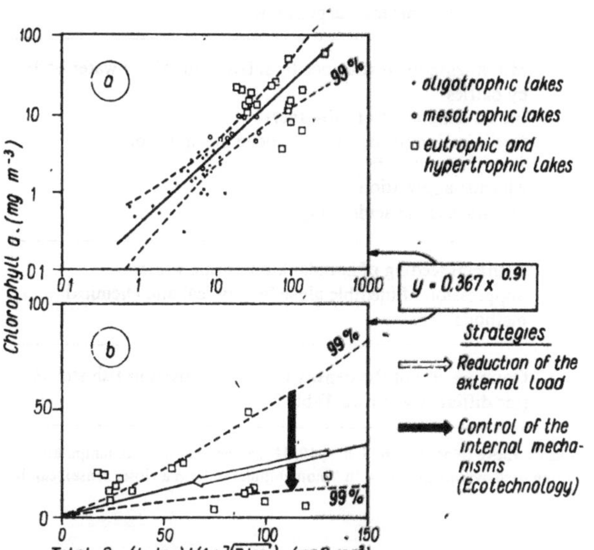

Fig. 1. The relationship between total phosphorus and chlorophyll a in the epilimnion of lakes in North America, (a) according to VOLLENWEIDER (1976). (b) transformation in a linear scale. The area between the 99%-confidence limits represents the control space which can be used by strategy (1) (white arrow) and strategy (2) (black arrow).

Lp — areal P-load (mg $P_{total} \cdot m^{-2} \cdot year^{-1}$), qs — hydraulic load ($m \cdot year^{-1}$), $\bar{z}$ — mean depth (m).

consequences of eutrophication or acidification and, therefore, it guarantees water use which otherwise would not be possible with the given external load.

Implementation of strategy (1) offers practically no unsolved scientific problems in eutrophication control (cf. VOLLENWEIDER 1968, BERNHARDT 1978, FORSBERG 1985). Consequently, the investigation, development and application of methods to implement strategy (2) are a field of increasing scientific and practical interest (cf. SHAPIRO et al. 1975, KLAPPER 1980, BENNDORF et al. 1981, UHLMANN 1983, BJÖRK 1985).

Among the some ten basic ecological control variables which can be employed in ecotechnology (Table 1, see also STRAŠKRABA 1979), the mortality of the herbivorous zooplankton is of particular limnological interest. All the other variables (Nos. 1 to 10 in Table 1) are used by physical, chemical or biological measures for controlling the trophic pyramid from the bottom to the top. Only zooplankton mortality provides the possibility of a top-down control of the community structure.

Many investigators were able to show experimentally that a maximization of the "filtration" capacity of the herbivorous zooplankton can be achieved by controlling their losses to predation caused by the higher trophic levels (carnivorous rotifers and crustaceans, insects, fish). The primary measure of all these investigations consists in creating a low (or more precisely: an optimum) density of zooplanktivorous fish by different methods (Table 2), and the main objective of this measure is to achieve the dominance of large herbivores (above all *Daphnia*), which can effectively reduce the abundance of certain kinds of phytoplankton. The most obvious effect with respect to water quality is the increase in the transparency. In some cases even a decrease in the nutrient concentration was caused by this top-down control (e.g. STENSON et al. 1978, HENRIKSON et al. 1970, SHAPIRO & WRIGHT 1984, WRIGHT & SHAPIRO 1984).

But there are some unsolved problems if food-web manipulation ("biomanipulation" according to SHAPIRO et al. 1975) is to be used as a measure to reduce the consequences of eutrophication:

(1) Development of such algae which are not eaten or digested by herbivores (STENSON et al. 1978, MCCAULEY & BRIAND 1979, NICHOLLS et al. 1980, LYNCH & SHAPIRO 1981, GLIWICZ 1981, BENNDORF et al. 1984, INFANTE & ABELLA 1985). Under which circumstances (nutrient supply, mixing conditions etc.) do blue-greens dominate among these inedible algae? What are the reasons of the development of toxic strains among blue-greens? Which positive and negative feedback mechanisms exist between dominance of different types of blue-greens and higher trophic levels (zooplankton, insects, fish)?

Table 1. Ecological control variables and practical measures for its use in ecotechnology (i.e. in-lake eutrophication control) (from BENNDORF et al. 1981, extended)

No.	Ecological control variable	Practical measures
Ia	Internal phosphorus load from the sediment: control by the redox potential	— Hypolimnion aeration — Introduction of pure oxygen — Enlargement of the hypolimnic oxygen pool by increasing the mean depth or introducing cool water, rich in oxygen — Injection of nitrate into the sediment (Riplox) — Introduction of effluents from tertiary swage treatment (poor in phosphorus; rich in nitrate) in the sediment-water interface
1b	Internal phosphorus load from the sediment: control by mechanisms not directly related to the redox potential	— Lake bottom sealing with clay minerals or aluminium sulfate — Creating a macrophyte cover on the sediment — Minimizing the stock of benthivorous fish which grub in the sediment — Sediment dredging
2	Phosphorus export by discharge	— Deep water release (reservoirs) — OLSZEWSKI-tube (lakes)
3	Phosphorus export into the sediment	— Spraying aluminium or iron minerals on the water surface — Addition of aluminium or iron minerals to the inflowing water — Stocking with silver carps (sedimentation of the fecal pellets) — Promoting the larger herbivorous zooplankton (vertical migration and sedimentation of the fecal material) — Optimization of the natural calcite precipitation
4	Temperature stratification and depth in which inflowing water is inserted	— Hypolimnic injection of outfalls from sewage treatment
5	Water residence time	— Flushing with water low in phosphorus
6a	Mixing depth; control of the light conditions	— Artificial destratification
6b	Mixing depth: control of the chemical conditions	— Artificial destratification
7	Light extinction	— Introducing water rich in humic substances — Introducing mineral suspensions
8	Phytoplankton sedimentation	— Minimizing the turbulence by subdivision of the water body by baffles — Introducing mineral substances — Optimization of the natural calcite precipitation
9	Phytoplankton mortality	— Algicide application — pH shock (lime acids, CO_2)
10	Zooplankton growth	— Artificial feeding of zooplankton during winter — Suppression of inedible algae by physical and chemical methods
11	Zooplankton mortality[1]	— Optimization of the density of zooplanktivirous fish stocks (for different ways see Table 2)

[1] According to SHAPIRO et al (1975), SHAPIRO (1979) many of the control variables listed above (Nos 1, 3, 6, 7, 8, 9, 10, 11) can be used in "biomanipulation" But it should be stressed that "top-down control of the foodweb", which is very often used synonymously with "biomanipulation" (in a closer sense), can be achieved exclusively by using variable No 11.

(2) The structure of a zooplankton community is governed not only by predation but simultaneously by competition for food (cf. LAMPERT 1985). The same cooperative effects occur in other communities (insects, fish. cf. LESSMARK 1983, NEILL 1985). What are the indirect effects of this interaction of predation and competition if one of the factors (predation) is changed artificially? Since some species of interest have generation times of one year and longer, these effects can be investigated in a long-term scale only.

(3) Which roles are played by spatial inhomogeneity of the physical and chemical conditions in controlling the community structure (TIMMES 1984, HORNIG & BENNDORF 1985)?

(4) What is the upper limit of the external and internal phosphorus load which should not be exceeded if

Table 2 Methods used in biomanipulation (control of a desirable structure of the food-web from the top of the trophic pyramid downwards)

Method	Experimental water body	Authors
Enclosures	Lake Emily	SHAPIRO et al (1975)
	Lakes Trummen and Bysjon	ANDERSSON et al (1978)
	Green Lake	GOAD (1984)
Experimental poisoning	Elbe backwaters	HRBAČEK et al (1961)
	Lake Lilla Stockelidsvatten	STENSON et al (1978)
		HENRIKSON et al (1980)
	Round Lake	SHAPIRO & WRIGHT (1984)
	Pond Habo 1	ANDERSSON & CRONBERG (1984)
	Wirth Lake	SHAPIRO (1982)
	Lake Haugatjern	REINERTSEN & OLSEN (1984)
Partial poisoning of an artificial divided lake	Pleasant Pond	LYNCH (1979)
		LYNCH & SHAPIRO (1981)
Enhancement of the stock of piscivorous fish	Hubenov reservoir	HRBAČEK et al. (1978)
	Lake Trummen	ANDERSSON et al. (1978), BJÖRK (1978)
	Loch Loso	LYNCH & SHAPIRO (1982)
	Experimental water Gräfenhain	BENNDORF et al (1984)
	Bautzen reservoir	BENNDORF (1979)
		HORNIG & BENNDORF (1985)
		BENNDORF et al (1988)
	Wirth Lake	SHAPIRO (1982)
	Round Lake	SHAPIRO & WRIGHT (1984)
		WRIGHT & SHAPIRO (1984)
	Lake Lenore	EDMONDSON & LITT (1984)
	Feldberg Haussee	KASPRZAK et al (1988)
Selective catch and/or catch restrictions	Lake Trummen	ANDERSSON et al. (1978)
		BJÖRK (1978)
	Bautzen reservoir	BENNDORF et al. (1988)
	Feldberg Haussee	KASPRZAK et al. (1988)
Improvement of the conditions of reproduction	Lake Washington with Cedar river[1]	EDMONDSON & LITT (1982)
		EDMONDSON & ABELLA (1988)

[1] The method was not used intentionally but hitherto there is no other example

food-web control is to operate successfully? Decreasing in-lake phosphorus concentrations have been explained by WRIGHT & SHAPIRO (1984) as an increase in the zooplankton-induced P-losses. Is this mechanism an indispensable prerequisite for the long-term success and stability in biomanipulation?

(5) Is the extermination of zooplanktivorous fish by poisoning the whole lake an appropriate way to minimize predation on herbivorous zooplankton in a long-term scale? As predation on large carnivorous zooplankton also is automatically minimized in this way, the final result of such an extermination of fish could be a rather low density of herbivores due to large populations of invertebrate predators. Thus, other methods of biomanipulation should be sought and tested, to reduce the zooplanktivorous fish but still guarantee a stable low stock of these fish, able to control the invertebrate predators. Is stocking with predacious fish an appropriate measure to achieve this goal? To what extent can the approach be generalized to improve the conditions of reproduction, growth of survival of the desirable fish (EDMONDSON & ABELLA 1988)?

The aim of this volume is to try some answers to these questions.

References

ANDERSSON, G, BERGGREN, H, CRONBERG, G, & GELIN, C (1978) Effects of plantivorous and benthivorous fish on organisms and water chemistry in eutrophic lakes. Hydrobiologia 59, 9–15

— & CRONBERG, G (1984) *Aphanizomenon flos-aquae* and large *Daphnia* — An interesting plankton association in hypertrophic lakes In BOSHEIM, S., & NICHOLLS, M (ed.), Interactions between trophic levels in freshwaters, Oslo, 63–76

BENNDORF, J (1979) Kausalanalyse, theoretische Synthese und Simulation des Eutrophierungsprozesses in stehenden und gestauten Gewässern Diss. B, Techn. Univ Dresden, Fakultät f Bau-, Wasser- u. Forstwesen, 1–165

— UHLMANN, D., & PÜTZ, K. (1981) Strategies for water quality management in reservoirs in the German Democratic Republic. Water Quality Bulletin 6 (3), 68–73 and 90

— KNESCHKE, H, KOSSATZ, K., & PENZ, E (1984) Manipulation of the pelagic food web by stocking with predacious fishes Int Rev ges. Hydrobiol 69, 407–428

— SCHULTZ, H., BENNDORF, A., UNGER, R., PENZ, E, KNESCHKE, H, KOSSATZ, K., DUMKE, R., HORNIG, U, KRUSPE, R., & REICHEL, S. (1988) Food-web manipulation by enhancement of piscivorous fish stocks Long-term effects in the hypertrophic Bautzen reservoir, *Limnologica* 19 (1), 97–110.

BERNHARDT, H (ed.), (1978) Phosphor – Wege und Verbleib in der Bundesrepublik Deutschland Weinheim, 1–285

BJÖRK, S. (1978) Restoration of degraded lake ecosystems Proceedings of the MAB Project 5 Regional Workshop "Land use impacts on lake and reservoir ecosystems", Warsaw, May 26 to June 2, 196–219

— (1985): Lake restoration techniques. Proceedings preprints of the International Congress "Lakes pollution and recovery", Rome, 15th—18th April, 202—212.

EDMONDSON, W. T., & ABELLA, S. E.B. (1988): Unplanned biomanipulation in Lake Washington, *Limnologica* **19** (1), 73—79.

— & LITT, A. H. (1982): *Daphnia* in Lake Washington. Limnol. Oceanogr **27**, 272—293.

— — (1984): Mt. St. Helens ash in lakes in the lower Grand Coulee, Washington State, Verh. Int. Verein. Limnol. **22**, 510—512.

FORSBERG, C. (1985): Lake recovery in Sweden. Proceedings preprints of the International Congress "Lakes pollution and recovery", Rome, 15th—18th April, 272—281.

GLIWICZ, Z M. (1981): The phase of summer increase in phytoplankton biomass. — Plankton Ecological Group Seasonal Succession Project, Draft to be discussed at PEG meeting in Constance, 1—6.

GOAD, J A. (1984): A biomanipulation experiment in Green Lake, Seattle, Washington. Arch. Hydrobiol. **102** (2), 137—153.

HENRIKSON, L., NYMAN, H. G., OSCARSON, H. G., & STENSON, J. A. (1980): Trophic changes, without changes in the external nutrient loading. Hydrobiologia **68**, 257—263.

HORNIG, U., & BENNDORF, J. (1985): Biomanipulation in the Bautzen reservoir: spring zooplankton succession and consequences of spatial inhomogeneity in fish predation on summer zooplankton, J. Hydrobiol. **1**, 53—61.

HRBAČECK, J., DVOŘAKOVA, M., KOŘINEK, V., & PROCHAZKOVA, L. (1961): Demonstration of the effect of fish stock on the species composition and the intensity of metabolism of the whole plankton association. Verh. Int. Verein. Limnol. **14**, 192—195.

— DESORTOVA, B., & POPOVSKY, J. (1978): The influence of the fishstock on the phosphorus-chlorophyll ratio. Verh. Int. Verein. Limnol. **20**, 1624—1628.

INFANTE, A., & ABELLA, S. E. B. (1985): Inhibition of *Daphnia* by *Oscillatoria* in Lake Washington. Limnol. Oceanogr. **30**, 1046—1052.

KASPRZAK, P., BENNDORF, J., KOSCHEL, R., & RECKNAGEL, F. (1988): Applicability of the Food-web Manipulation in the Restoration Program of a Hypertrophic Stratified Lake: Model Studies for Lake Haussee (Feldberg, GDR). *Limnologica* (Berlin) 19 (1), 87—95.

KLAPPER, H. (1980): Experience with lake and reservoir restoration techniques in the German Democratic Republic. Hydrobiologia **72**, 31—41.

LAMPERT, W. (ed.) (1985): Food limitation and the structure of zooplankton communities. Arch. Hydrobiol., Beih. Ergebn. Limnol. **21**, 497 pp.

LESSMARK, O. (1983): Competition between perch (*Perca fluviatilis*) and roach (*Rutilis rutilus*) in south Swedish lakes. Doct. Diss., Lund Univ., Inst. of Limnology, 1—172.

LYNCH, M. (1979): Predation, competition and zooplankton community structure: An experimental study. Limnol. Oceanogr. **24**, 253—272.

— & SHAPIRO, J. (1981): Predation, enrichment and phytoplankton community structure. Limnol. Oceanogr. **26**, 86—102.

— — (1982): Manipulations of planktivorous fish — Effects on zooplankton and phytoplankton. In: SHAPIRO, J., FORSBERG, B., LAMARRA, V., LINDMARK, G., LYNCH, M., SMELTZER, E., & ZOTO, G., Experiments and Experiences in Biomanipulation. Interim Report No. 19, Limnological Research Center, Univ. of Minnesota, Minneapolis, Minnesota, 158—188.

MCCAULEY, E., & BRIAND, F. (1979): Zooplankton grazing and phytoplankton species richness: field tests of the predation hypothesis. Limnol. Oceanogr. **24**, 243—252.

NEILL, W. E. (1985): The effects of herbivore competition upon the dynamics of *Chaoborus* predation. Arch. Hydrobiol., Beih. Ergebn. Limnol. **21**, 483—491.

NICHOLLS, K. H., KENNEDY, W, & HAMMETT, C. (1980): A fish-kill in Heart Lake, Ontario, associated with the collapse of a massive population of *Ceratium hirundinella* (Dinophyceae). Freshwater Biology **10** (6), 553—561.

REINERTSEN, H., & OLSEN, Y. (1984): Effects of fish elimination on the phytoplankton community of a eutrophic lake. Verh. Int. Verein. Limnol. **22**, 649—657.

SHAPIRO, J. (1979). The need for more biology in lake restoration. In. Lake restoration, Proceedings of a National Conference. Minneapolis, U.S. EPA Report No. 44/5-79-001, 161—167.

— (1982): Effects of rotenone in Wirth Lake. In: SHAPIRO, J., FORSBERG, B., LAMARRA, V., LINDMARK, G., LYNCH, M., SMELTZER, E., & ZOTO, G., Experiments and Experiences in Biomanipulation. Interim Report No. 19 of the Limnological Research Center, Univ. of Minnesota, Minneapolis, 195—207.

— LAMARRA, V., & LYNCH, M (1975): Biomanipulation: an ecosystem approach to lake restoration. In: BREZONIK, P. L., & FOX, J. L. (ed.), Water Quality Management through Biological Control. Report No. ENV-07-75-1, University of Florida, Gainesville, 85—96.

— & WRIGHT, D. I. (1984): Lake restoration by biomanipulation: Round Lake, Minnesota, the first two years. Freshwater Biology **14**, 371—383.

STENSON, J. A. E., BOHLIN, T., HENRIKSON, L., NILSSON, B. I., NYMAN, H. G., OSCARSON, H. G., & LARSSON, P. (1978): Effects of fish removal from a small lake. Verh. Int. Verein. Limnol. **20**, 794—801.

STRAŠKRABA, M. (1979): Mathematische Stimulation der Produktionsdynamik in Gewässern und deren Anwendung auf die Produktionssteuerung in Talsperren. Z. f. Wasser- und Abwasserforschung **12**, 56—64.

TIMMES, R. M. (1984): Prevention of growth of potentially dense phytoplankton populations by zooplankton grazing in the presence of zooplanktivorous fish in a shallow wetland ecosystem. Limnol. Oceanogr. **29**, 472—486.

UHLMANN, D. (1983): Entwicklungstendenzen der Ökotechnologie. Wiss. Z. Techn. Univ. Dresden **32** (6), 109—116.

VOLLENWEIDER, R. A. (1968): The scientific basis of lake and stream eutrophication, with particular reference to phosphorus and nitrogen as eutrophication factors. Tech. Rep. OECD, Paris, DAS/CSI 68, 27, 1—182.

— (1976): Advances in defining critical loading levels for phosphorus in lake eutrophication. Mem. Ist. Ital. Idrobiol. **33**, 53—83.

WRIGHT, D. I., & SHAPIRO, J. (1984): Nutrient reduction by biomanipulation: an unexpected phenomenon, and its possible cause. Verh. Int. Verein. Limnol. **22** (1), 518—524.

Dresden, February 1987 JÜRGEN BENNDORF

Part I

Interactions between Particular Components of the Food-web

Limnologica (Berlin) 19 (1), 11–20, Juli 1988

Max-Planck-Institut für Limnologie, Plön, FRG

The Relationship between Zooplankton Biomass and Grazing: A Review*

By WINFRIED LAMPERT

With 8 Figures and 3 Tables

Key words Zooplankton-phytoplankton interactions, Community grazing, Particle elimination, Clear-water phase, Weight-specific clearance rate, Biomass-grazing relationship

Abstract

Because of their central position in food webs, zooplankton are a key factor for successful lake management by "biomanipulation" This paper reviews the available literature to address the question of how much zooplankton is needed to control seston concentrations effectively During periods when zooplankton build up biomass between 1 5 and 4 0 mg dr w. m^{-2} due to both favourable food conditions and low predation pressure, clear-water phases occur The connection between the seasonal cycle of Secchi transparency and grazing activity is demonstrated by a particular example (Schöhsee). Grazing rates measured in situ may be as high as 200%/day in mesotrophic lakes but are much lower in eutrophic systems High grazing rates, however, may not always mean high rates of particle elimination. Thus, in this paper the amount of carbon eliminated by a given biomass of zooplankton is estimated from the metabolic requirements of the grazers. This relationship is then used to estimate the zooplankton biomass needed to offset a range of primary productivities of nannoplankton.

Introduction

In recent years, numerous studies have shown that the experimental manipulation of planktonic communities can help to restore eutrophied lakes (ANDERSON et al. 1978, HENRIKSON et al. 1980, SHAPIRO 1980, SHAPIRO et al. 1982, SHAPIRO & WRIGHT 1984, REINERTSEN & OLSEN 1984, BENNDORF et al. 1984, CARPENTER et al. 1985). These manipulations are based on increasing the biomass of large filter-feeding zooplankton and thus creating heavy grazing pressure on the phytoplankton. Increased grazing removes nannoplanktonic algae, bacteria and small detrital particles from the water. Many examples (for reviews see HALL et al. 1976, ZARET 1980, GREENE 1983) demonstrate that large zooplankton species dominate planktonic communities in the absence of planktivorous fish, as predicted by HRBÁČEK (1962) and BROOKS & DODSON (1965) The elimination of fish from a lake community often results in a rapid increase of zooplankton body size (e.g. SHAPIRO & WRIGHT 1984). The expected "cascading effect" on phytoplankton, however, is not always clearly seen (THRELKELD 1987, MCQUEEN et al. 1986), so that the debate over "bottom up" or "top down" control of phytoplankton continues.

A complete elimination of planktivorous fish will not always be possible. Poisoning by rotenone is not always applicable and, moreover, probably not justified from an ethical standpoint. There may be other technical and political reasons why planktivorous fish cannot be completely excluded. Moreover, even if planktivorous fish have been eliminated, it is difficult to maintain a lake free of fish for many years. Thus, on the long run it may be more fruitful to reduce planktivorous fish populations by continuing intensive fishing or by stocking with piscivorous fish, as practiced successfully in the GDR (BENNDORF et al. 1984). Under these conditions, there will always be some planktivorous fish present, but at low abundance Thus, zooplankton will always suffer some mortality from fish predation and will not reach maximum biomass. Therefore, it is of practical interest to develop a predictive theory of the zooplankton biomass required to produce a desired improvement of the water quality and the reduction in the fish stock necessary to produce this standing crop of zooplankton.

The purpose of this paper is to establish a zooplankton — grazing relationship from field measurements to allow prediction of how much zooplankton is needed to significantly reduce the particulate matter in a particular lake

Theoretical Considerations

Components of net change

As summarized by REYNOLDS (1984), the net change of phytoplankton biomass can be described by

$$\frac{dB}{dt} = B_t\ (k' - k_d - k_s - k_g) \quad (1)$$

i.e., by the biomass at time $t(B_t)$ and the instantaneous rate constants of growth (k'), death and decomposition (k_d) sedimentation (k_s), and grazing (k_g). The per capita rate change of biomass (k_n) is

$$k_n = k' - (k_d + k_s + k_g) \quad (2)$$

Biomass increases if k_n is positive but decreases if k_n is negative; no change is observed when $k_n = 0$.

Since the aim of biomanipulation is to reduce algal biomass, k_n should be made negative Thus the sum of all the loss rate variables (in parentheses) should first be greater than, and then equal k'. The rate variables k' and k_g can be influenced by manipulating zooplankton An increase of herbivorous zooplankton biomass will increase k_n but may, on the other hand, enhance k' by increased regeneration of limiting nutrients (LEHMAN 1980, STERNER 1986). However, k_d and k_s can never be negative and any increase in these two rate variables will always expedite the biomass elimination. We will, therefore, get a reduction of particulate biomass anytime $k_n > k'$

Even in lakes with fish, cases have been reported where the phytoplankton losses were nearly exclusively caused

* An earlier version of this article was presented at the EWPCA International Congress „Lakes Pollution and Recovery", Rome, April 1985

by zooplankton grazing at certain times of the year (GLIWICZ & HILLBRICHT-ILKOWSKA 1975, REYNOLDS et al. 1982, TILZER 1984). SOMMER (1984) showed that the changing algal abundance during spring in Lake Constance could be explained by only two factors, light (controlling k') and zooplankton (determining k_g).

Estimates of the growth rate

Values of k' can be estimated from the daily primary production and the biomass (TILZER 1984). They can be as high as $1.5 \cdot d^{-1}$, but usually are much lower (REYNOLDS 1984).

Such estimates are often based on live phytoplankton biomass as calculated from algal volume. Since the mass of all particles, including detritus, is to be reduced by the manipulation, k' should be defined as the instantaneous growth rate of the whole particulate carbon pool. This can be done by considering POC as the basis for calculation, which makes the calculation even easier. Depending on the share of detritus, the growth rate of total particulate carbon is always lower than the algal growth rate, if resuspension from sediments can be neglegted.

Estimates of grazing rate

Several methods are available to estimate k_g. They may be based on in situ measurements (GLIWICZ 1968, HANEY 1973), or collecting animals and treating them under in situ conditions (GULATI 1975, HORN 1981, THOMPSON et al. 1982). The decrease of particles in enclosed volumes of water (GLIWICZ 1968, PORTER 1973, HORN 1981, CRUMPTON & WETZEL 1982) has been measured, as well as the uptake of radioactively labelled particles (HANEY 1973, LAMPERT & TAYLOR 1985, THOMPSON et al. 1982), or labelled natural algal assemblages (GULATI 1975, ROSS & MUNAWAR 1981, OKAMOTO 1984). These methods determine the fraction of water cleared of the particles under consideration per unit time. This fraction is numerically equivalent to k_g. None of them, however, can be used to correctly estimate the loss rate of all particles. Cell-count methods do not consider detritus. Radiotracer techniques measure the uptake of only the labelled particles. Material ingested is not always removed from the water, since some particles may pass the gut undigested (PORTER 1973) or may be converted into detritus. Methods using labelled tracer algae assume that all particles are retained with the same efficiency as the tracer. This is certainly not the case (KNISELY & GELLER 1986). Thus clearance rates (% of water cleared per day) determined in this way by use of a well ingested particle will always overestimate the rate of removal of all particles; they must be multiplied by an efficiency term (REYNOLDS et al. 1982, VANDERPLOEG et al. 1984) for every combination of zooplankton and seston composition.

Estimates of the rate of particle removal

While in situ grazing measurements may provide an estimate of maximum particle removal rate, a minimum estimate can be obtained from the zooplankton biomass alone. The energy balance of zooplankton can be formulated as

$$P = I - F - E - R \tag{3}$$

where: P is the accumulated energy (production), I is the energy ingested, F and E are losses by defecation and excretion, respectively, and R denotes the respiratory loss.

Assuming that the excreta (E) are not removed from the water since they provide the substrate for the new growth of heterotrophic particles, the amount of energy removed (Z) is

$$Z = I - (F + E) \tag{4}$$

or

$$P = Z - R \tag{5}$$

Therefore, the amount of energy (or mass) eliminated can be estimated from

$$Z = R + P \tag{6}$$

Since a zooplankton population cannot persist without any production, the minimum estimate of Z would result from $P = 0$:

$$Z_{min} = R$$

Thus the minimum amount of particles eliminated can be estimated from the respiratory rate of the community of herbivorus zooplankton, an approach used by NOWAK (1975).

In fact, when integrated over a longer period of time, P will always be greater than zero. Unfortunately, secondary production is difficult to estimate and varies with the environmental conditions, especially with the availability of food. However, it has been shown that the coefficient of utilization of assimilated food, K_2, is a rather useful tool for estimating the seasonal average production (WINBERG 1971).

$$K_2 = (A - R)/A = P/A \tag{7}$$

where A is the amount of energy assimilated.

If we consider E (Equ. 4) to be small, A can be replaced by Z:

$$K_2 \approx P/Z; \quad P \approx K_2 \cdot Z \tag{8}$$

We obtain a second, more realistic, estimate of Z by substituting Equ. 8 into Equ. 6:

$$Z \approx R/(1 - K_2) \tag{9}$$

Field estimates of K_2 vary between 0.6 and 0.2, the average being approximately 0.3 (WINBERG 1971). Since the respiratory rate can be estimated from the biomass of herbivores (LAMPERT 1984), Equ. 9 provides a means of estimating the biomass of zooplankton required to eliminate a certain amount of particulates per time. Note that Z is the quantity eliminated completely via respiration; it does not represent the quantity ingested and transformed into a different component of the particulate carbon pool.

Field Evidence

Periods when phytoplankton is reduced by zooplankton, resulting in very clear water, have been described for sewage ponds (UHLMANN 1971), low-stock fish ponds (FOTT et al. 1980a, b), and also for a number of meso-

and eutrophic lakes (LAMPERT 1978, OSKAM 1978, BÜRGI et al. 1985, LAMPERT et al. 1986, SOMMER et al. 1985). Provided that the sampling frequency is high enough, a period of extremely clear water following the spring bloom of small algae can be detected in many meso- and eutrophic lakes. Several seasonal records of phytoplankton biomass show clear spring depressions (e.g. BLAAUBOER 1982, BAILEY-WATTS 1982, RIEMANN 1983, BARKO et al. 1984, MEFFERT & OVERBECK 1985) but do not mention this explicitly. A well-studied example is Lake Constance, where the Secchi transparency may increase from a little over 1 m to more than 10 m within a few days (LAMPERT 1978). This seasonal event is very predictable (GELLER 1980). Clearwater phases usually coincide with spring maxima of zooplankton (Fig. 1). Under low fish predation pressure, before the new year

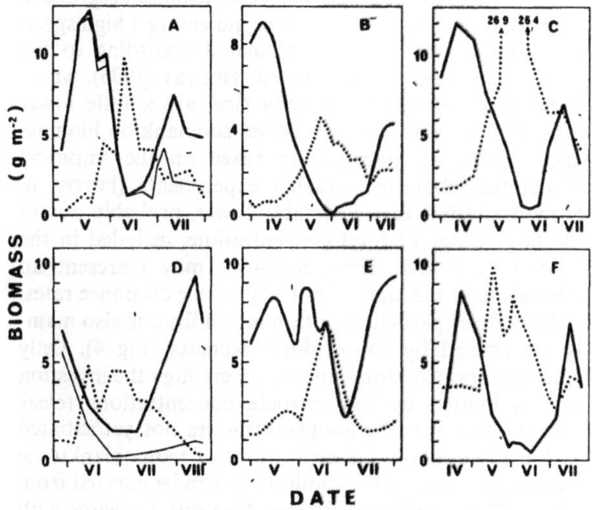

Fig. 1. Six examples of zooplankton induced clear-water phases.

Thick solid line: phytoplankton biomass (carbon); thin line: biomass of phytoplankton <35 µm (carbon), dotted line: zooplankton biomass (dry weight).

A. Lake Constance 1977 (SCHOBER 1980), B: Plußsee 1969 (KRAMBECK et al. 1978); C: Greifensee 1977 (BÜRGI et al 1985); D: Lake Esrom 1974 (PETERSEN 1983), E. Lake Biel 1978 (TSCHUMI et al. 1982); F: Lake Lucerne (BÜRGI et al 1985)

class of fish starts foraging, and with excess food, zooplankton populations increase exponentially. The weight-specific grazing activity may increase at the same time because of the increasing temperature. Thus k_g exceeds k' and, as a consequence, the water transparency improves.

When zooplankton populations crash after the spring peak, algal biomasses usually recover, but the new biomass is often composed of different algal species. In examples Fig. 1A and Fig. 1D particulates have been fractionated into size groups. These examples show that the recovery of phytoplankton biomass after the zooplankton peak is mainly due to the growth of large particles, while small particles remain suppressed at even lower zooplankton densities during summer. The figure illustrates the seasonal shift from small "edible" algae to larger "ungrazable" phytoplankton (SOMMER et al. 1985).

Unfortunately, many studies describing the time course of phytoplankton parameters do not give the zooplankton biomass simultaneously; most of them report only numbers. Only a small selection can, therefore, be used to estimate the zooplankton biomass present when the clear-water phase is initiated. A compilation of such "critical" zooplankton densities from the literature is presented in Table 1. Data are gathered from lakes of different trophic state. "Critical" concentrations of zooplankton are read from the published figures at the points where significant losses of phytoplankton biomass or increases of the Secchi transparency were observed. Therefore, they cannot be very accurate. Moreover, the decision at which point of the seasonal curve "significant losses" led to a clearwater phase is somewhat subjective; hence, the data compiled in Table 1 are rough estimates. Despite of these potential errors, the "critical" concentrations are surprisingly similar. Although they tend to increase with lake trophy, the majority of values are between 1.5 and 4.0 mg dr.w. · m^{-2} of zooplankton. Bieler See in Switzerland is a little exceptional in that the zooplankton maximum preceeds the phytoplankton minimum (cf. Fig. 1E). The reason may be that the zooplankton biomass in this lake is mainly composed of rotifers, while crustaceans dominate in all the other lakes.

Table 1. Estimates of the "critical" concentration of zooplankton at times when significant losses of particles due to grazing are to be observed during seasonal succession. Sources: 1 — CRUMPTON & WETZEL (1982); 2 — REINERTSEN & LANGELAND (1982); 3 — DEBERNARDI (1974); 4 — HORN (1981); 5 — WRIGHT (1958); 6 — SCHOBER (1980); 7 — LAMPERT (1978); 8 — BÜRGI et al. (1985); 9 — LAMPERT et al. (1986); 10 — KRAMBECK et al. (1978); 11 — NOWAK (1975); 12 — STEEL et al. (1972), 13 — TSCHUMI et al. (1982)

Lake	Trophic state	"Critical" concentration of zooplankton		Seasonal zooplankton maximum	Source
		Date	(g dr.w. · m^{-2})	(g dr.w. · m^{-2})	
Lawrence Lake	oligo	8/10/79	1.9	2.4	(1)
Langvatn	oligo	7/03/75	1.0	4.7	(2)
Lago Maggiore	meso	5/28/73	3.1	5 4	(3)
Saidenbach Res.	meso	5/01/75	1.5	7.0	(4)
Canyon Ferry Res.	meso	9/11/56	3.6	6.5	(5)
Bodensee	meso	5/20/77	2.0	9.8	(6)
Bodensee	meso	5/15/76	2.4	4.0	(7)
Lake Lucerne	meso	5/10/79	4.5	9 8	(8)
Schöhsee	meso-eu	5/05/83	2.5	8.8	(9)
Plußsee	eu	5/01/69	1.5	4 6	(10)
Schierensee	eu	6/01/69	3.0	9.0	(11)
Queen Mary Res.	eu	4/15/68	3.0	12.0	(12)
Greifensee	eu	5/01/77	4.0	26.9	(8)
Bieler See	eu	6/10/78	4.5	6.9	(13)

In Situ Grazing Rates

Seasonal cycle of grazing

The relative importance of growth and loss processes for the dynamics of algal populations varies during the season. A detailed seasonal cycle of primary production and grazing rates can be constructed from the studies on the moderately eutrophic Schöhsee in the northern part of FRG (LAMPERT & TAYLOR 1984, 1985, LAMPERT et al. 1986). Daily average grazing rates on nannoplankton in the upper 5 m of the lake, measured in situ, peak in the middle of May with 170% per day. The peak coincides with the zooplankton maximum and the spring clear-water phase. Daily primary production has been opposed to grazing losses (Fig. 2). This is easy in this case (TILZER 1984), since photosynthesis takes place during the day while most of the grazing losses occur at night (HANEY & HALL 1975, LAMPERT & TAYLOR 1985). The amount of carbon removed from the water has been calculated by multiplying the average daily grazing rate (% of water cleared per day) with the concentration of sestonic particles smaller than 35 µm. For the reasons discussed above, it was assumed that 50% of the ingested carbon had been eliminated (cf. GULATI et al. 1982). Fig. 2 illustrates that concentrations of particulate carbon (<35 µm) below 0.3 mg · l^{-1} and Secchi transparencies of more than 5 m occur during periods when the daily rate of elimination exceeds the daily rate of primary production, i.e., during May and September/October. It also shows that only the small fraction of seston (<35 µm) is affected by the elevated grazing pressure, larger particles (35—250 µm) are not.

Community grazing rates during the seasonal cycle in Schöhsee are strongly correlated with the abundance of herbivorous zooplankton (Fig. 3). This has also been found by HART (1986). Zooplankton body mass is the governing factor when in situ filtering rates in different lakes are compared by regression analysis (CHOW-FRASER & KNOECHEL 1985, KNOECHEL & HOLTBY 1986) or when all available laboratory feeding experiments are pooled (PETERS & DOWNING 1984). Other environmental factors like concentration of food and temperature play a relatively minor role. The slope of the Schöhsee regression (Fig. 3), however, is rather steep, indicating a high specific clearance rate. Values calculated according to the predictive model of KNOECHEL & HOLTBY (1986), which is based on zooplankton body size, are a little lower (Fig. 3). The relationship between zooplankton biomass and community grazing rate based on the empirical model for laboratory grazing experiments (PETERS & DOWNING 1984) is considerably lower, probably due to the large range of food concentrations included in the laboratory model. Thus, Schöhsee may represent an estimate near the higher limit of possible clearance rates.

The rate of particle ingestion in Schöhsee is also mainly determined by zooplankton biomass (Fig. 4). Only when the zooplankton biomass is very high the ingestion rate is limited by low particle concentrations (clear water). However, the zooplankton are not yet satiated within the range of concentrations of nannoplanktonic particles present in the Schöhsee, as can be inferred from Fig. 5. The weight-specific ingestion rate increases with the particle concentration and does not reach a plateau.

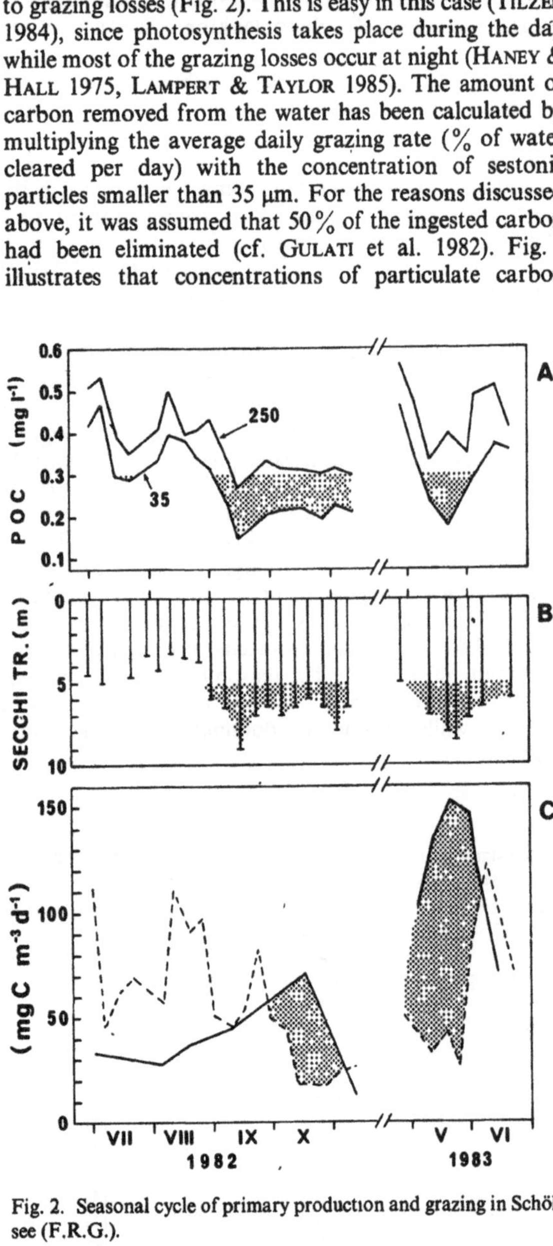

Fig. 2. Seasonal cycle of primary production and grazing in Schöhsee (F.R.G.).

A · Particulate carbon in the size fractions <35 µm and <250 µm
B Secchi transparency. C Average primary production (dashed line) and estimated average rate of particle elimination by zooplankton (solid line) in the upper 5 m of the lake. Rates of particle elimination were assumed to be 50% of the grazing rates measured in situ, multiplied by the concentration of seston (<35 µm). The shaded areas indicate the time periods when the rate of elimination exceeds the rate of production, Secchi transparency exceeds 5 m, and particulate carbon (<35 µm) is lower than 0.3 mg/l

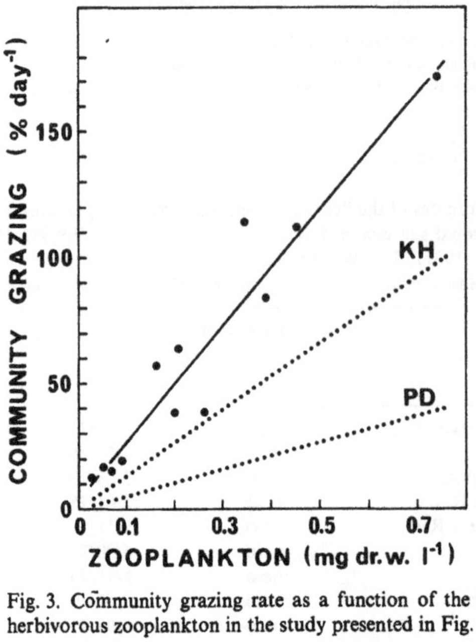

Fig. 3. Community grazing rate as a function of the biomass of herbivorous zooplankton in the study presented in Fig. 2.

Points represent average daily rates from measurements spread all over the season. The regression is $Y = 4.5 + 231 \cdot X$ ($r = 0.955$). The dotted lines are calculated according to the empirical regression for in situ grazing (KNOECHEL & HOLTBY 1986, KH) and laboratory feeding experiments (PETERS & DOWNING 1984; PD). Both relationships were constructed for cladoceran communities of 1.2 mm average body size

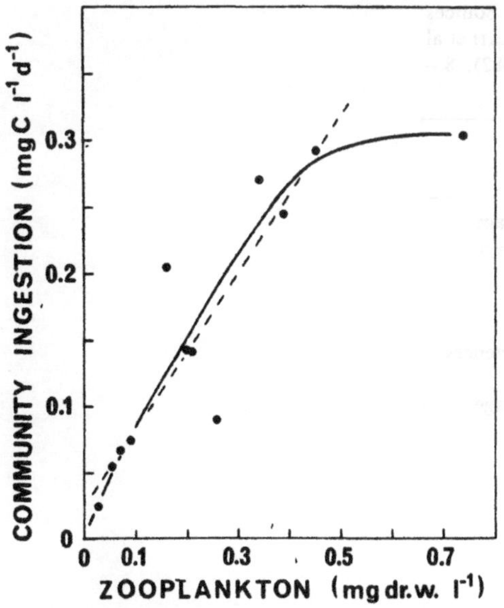

Fig. 4. Community ingestion rate in Schöhsee as a function of the biomass of herbivorous zooplankton.

The points correspond to Fig. 3. Solid line drawn by eye. Dashed line Regression including all points below 0 5 mg dr w l^{-1} ($Y = 0\ 027 + 0\ 583 \cdot X$, $r = 0\ 894$).

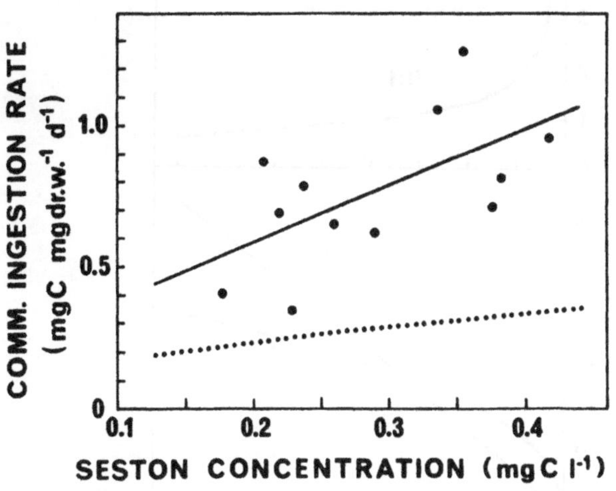

Fig. 5 Weight specific ingestion rate as a function of the seston (<35 μm) concentration.

Points correspond to Fig 4 The regression is $Y = 0.19 + 1\ 98\ X$ ($r = 0.604$) The dotted line represents weight specific ingestion rates calculated from the formula of GULATI (1983) for Dutch lakes (see text).

This figure also reveals that the ingestion rates calculated in this way (grazing rate measured by a single tracer, multiplied by the concentration of all particles <35 μm) are overestimates. At the high food concentrations they reach approx. 1 mg C · mg dry weight^{-1} · d^{-1} (equivalent to 240% per day on a carbon basis). Compared to laboratory experiments (PETERS & DOWNING 1984), these values appear unusually high. The in situ measurements were carried out with two particles simultaneously, the green alga *Scenedesmus* (10 μm), and the very small blue-green *Synechococcus* (1 μm), but only the grazing rates obtained with *Scenedemus* were used for the calculation of ingestion rates. Grazing rates for the small particle were consistently lower than those for *Scenedemus*, especially when size selective grazers like *Eudiaptomus* made up a considerable proportion of the zooplankton biomass. Thus we can assume that a great part of the particles within the 35 μm range is ingested with a lower efficiency than is *Scenedesmus*. Community ingestion rates calculated using the formula of GULATI (1983), which was derived from measurements with labelled seston, are, in fact, lower (Fig. 5).

Specific clearance rates

The in situ grazing rates reported in the literature are rather variable. They depend not only on the zooplankton biomass and on the environmental conditions, but also on the method applied. Table 2 lists the maximum community grazing rates obtained by different authors. Highest values were measured with the HANEY technique (HANEY 1973), lowest rates were measured with the cell count-method.

Table 2. Maximum community grazing rates measured in different lakes. Sources: 1 — HANEY (1973); 2 — HART (1984); 3 — LAMPERT & TAYLOR (1985); 4 — THOMPSON et al. (1982); 5 — WRIGHT (1958); 6 — GULATI et al. (1982); 7 — CRUMPTON & WETZEL (1982); 8 — HORN (1981)

Lake	Year	Maximum grazing rate (% per day)	Method	Source
Heart Lake	1968 1969	180 310	Haney chamber tracer, in situ	(1)
Lake Le Roux	1981	150*	Haney chamber tracer, in situ	(2)
Schöhsee	1983	170	Haney chamber tracer, in situ	(3)
Blelhem Tarn	1978	120	Tracer, laboratory	(4)
Canyon Ferry R.	1956	106	Estimate from phytopl. loss	(5)
Lake Vechten	1974 1978	40 60	Labelled natural seston	(6)
Lawrence Lake	1979	39	Tracer, beaker on board	(7)
Saiden-bach R.	1975 1976	28 13	cell count	(8)

* one value at 25 °C omitted

Because of the lack of biomass data, weight-specific clearance rates can only be calculated in a few cases (Table 3). Again the HANEY method yields the highest estimates (more than 2 1 · mg dr.w.$^{-1}$ · d^{-1}). However, the variation between lakes is not unexpected when we consider the differences of methods, the different species composition of the communities measured, the different concentrations and qualities of seston, and the different temperature conditions. Weight-specific clearance rates measured in the laboratory may even vary by more than two orders of magnitude (PETERS & DOWNING 1984). Average in situ rates of 1 1 · mg dr.w^{-1} · d^{-1} may be reasonable and are consistent with the critical concentrations of zooplankton listed in Table 1. KNOECHEL & HOLTBY (1986) point out that the exponents of the length-size relationship and of the relationship between size and in situ clearance rate are very similar, so that the grazing rate may be predicted from the zooplankton biomass. The specific clearance rate obtained by their model is well in the range of other estimates (Table 3).

Table 3 Examples of weight specific community clearance rates from the literature. Sources: 1 — LAMPERT & TAYLOR (1984); 2 — HART (1974); 3 — GLIWICZ (1968); 4 — GULATI et al. (1982); 5 — GULATI (1984); 6 — GULATI (1983); 7 — CRUMPTON & WETZEL (1982); 8 — KNOECHEL & HOLTBY (1986)

Lake	Source	Specific clearance rate (ml · mg dr.w.$^{-1}$ · d^{-1})	Comment
Schöhsee	(1)	2740	Seasonal average
Lake Le Roux	(2)	2345	Seasonal average
Mikolajkie	(3)	480/1045	July/August
Taltowisko	(3)	370/920	July/August
Lake Vechten	(4)	85—810	Seasonal differences
5 eutrophic Dutch lakes	(5)	143—243	Seasonal average
Oligotrophic Dutch lakes	(6)	1800	Average
Mesotrophic Dutch lakes	(6)	1200	Average
Lawrence Lake	(7)	660	At maximum grazing rate
Lake St. George	(8)	1320	From regression (cf. Fig. 3)

A comparison of different lakes

Probably the most thorough study of community grazing has been carried out by GULATI and his co-workers in Lake Vechten and a collection of Dutch lakes of varying trophy (GULATI 1975, GULATI et al. 1982, GULATI 1983, 1984). Although the method used has some disadvantages (for a critique see GULATI et al. 1982), it has been applied during several seasons and on many lakes in the same way, so that a comparison of average specific clearance rates for the different lakes is possible. Other than the studies mentioned earlier (CHOW-FRASER & KNOECHEL 1985, LAMPERT & TAYLOR 1985, HART 1986) the studies by GULATI and his collaborators cover a very wide range of seston concentrations and, thus, demonstrate the effect of seston concentration more clearly. Weight-specific clearance rates vary between 1.8 and 1.2 l · mg dr.w.$^{-1}$ · d^{-1} in oligotrophic and mesotrophic lakes (<1 mg C · l^{-1}) but decrease strongly in eu- and hypertrophic lakes. The relationship between the concentration of seston (C) smaller than 33 µm (in mg C · l^{-1}) and the specific clearance rate (SCR) (in ml · mg dr.w.$^{-1}$ · d^{-1}) can be described by

$$SCR = 533 \cdot C^{-0.487} \quad \text{(GULATI 1983).}$$

The weight-specific ingestion rate can be calculated from this formula by multiplication with the seston concentration.

A figure for the zooplankton biomasses corresponding to the above-mentioned specific clearance rates can be found in GULATI (1983). Average zooplankton biomasses in the very eutrophic lakes are lower than predicted by the empirical model of MCCAULEY & KALFF (1981). They are plotted versus the seston concentration in Fig. 6. Biomass increases rapidly with seston in the mesotrophic lakes but very slowly in the hypertrophic waters. The daily community ingestion rate (mg C · l^{-1} · d^{-1}) can be calculated by multiplication of the weight specific ingestion rate with the zooplankton biomass. Astonishingly, it increases in a nearly linear way with the seston concentration (Fig. 6):

$$I = 236 \cdot C^{0.915}$$

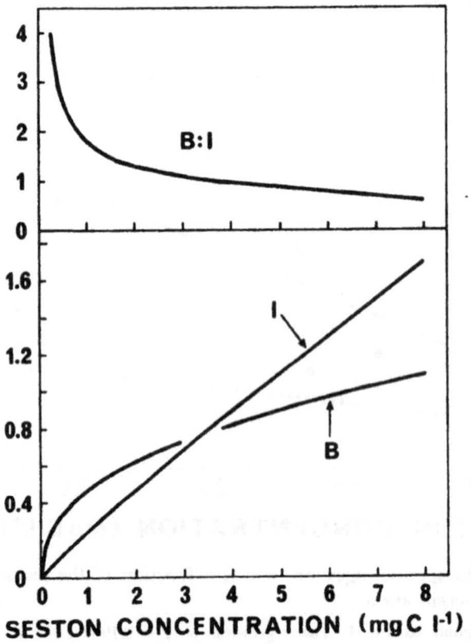

Fig. 6. Relationship of average zooplankton biomass and ingestion rate for particles <33 µm in Dutch lakes of varying trophy (calculated from GULATI 1983).

Lower panel: Zooplankton biomass (B) in mg dr.w · l^{-1} and specific ingestion rate (I) in mg C mg dr w^{-1} · d^{-1} as a function of seston concentration Upper panel: Ratio of B I.

The ratio B:I reflects the amount of biomass resulting from a given consumption of seston. It decreases strongly, indicating that at high seston concentrations a much higher ingestion is necessary to maintain a certain biomass than at low ones. There are two alternative explanations for this fact:

(1) The utilization of the ingested food for biomass production may become less and less efficient at high seston concentrations. One reason might be that the food passes through the gut too fast to be efficiently digested (superfluous feeding). Another possible explanation involves a gradual change of the seston

with the trophic state in such a way that the proportion of undigestible (PORTER 1973) or less valuable small particles increases with increasing total concentration.

(2) If the efficiency of food utilization for growth does not change, zooplankton production must increase according to the ingestion rate. Then the relatively small increase of the zooplankton biomass with increasing food concentration must be caused by intensive cropping of the produced biomass by predation.

The available data do not allow us to decide which of the alternatives is correct. However, the greatest change in the B:I ratio takes place at the lower range of food concentrations where the occurrence of "superfluous feeding" is unlikely (LAMPERT 1977). The assimilation efficiency of the natural food was rather constant throughout the season in Lake Vechten (GULATI et al. 1982). In the more eutrophic lakes, the assimilation efficiencies were smaller than in Lake Vechten (GULATI et al. 1985), but the B:I ratio (Fig. 6) covers a much greater range than explained by the difference in assimilation efficiencies. This seems to prove that the zooplankton biomass in the eutrophic lakes is controlled by predation rather than by food, a hypothesis supported by the empirical finding that the proportion of piscivorous fish decreases with lake trophy (L. PERSSON, pers. comm.).

Particle Elimination

The grazing rates reported in Table 2 are high enough to allow the control of phytoplankton through zooplankton. Under most conditions a rate of 80% per day ($k_g = 0.8$) should be sufficient to balance the growth rate of algae (REYNOLDS 1984). However, as pointed out earlier, this does not mean that 80% of all particles are eliminated per day. In the Dutch lakes (GULATI et al. 1985), the assimilation efficiencies for natural particles are often less than 30%, thus only a small amount of the ingested particles is eliminated from the seston.

Since the efficiency of particle elimination is difficult to determine, a physiological approach based on the carbon requirements of the zooplankton may be more profitable. The respiratory carbon losses of the most common species of herbivorous zooplankton are about 22% of the body carbon per day at 20 °C, with some variation around this figure (LAMPERT 1984). Corrected to 15 °C according to Krogh's curve (WINBERG 1971) respiratory losses amount to 14.5% per day. This is in good agreement with the field measurements of NOWAK (1975) (13.8% per day). Assuming 40% carbon in dry weight of zooplankton we would obtain (cf. Equ. 9):

$$R = 0.058 \text{ mg C} \cdot \text{mg dr.w.}^{-1} \cdot \text{d}^{-1}$$

By use of an average K_2 of 0.3 (WINBERG 1971) the daily amount of carbon eliminated can be calculated as

$$Z = 83 \text{ mg C} \cdot \text{g dr.w.}^{-1} \cdot \text{d}^{-1}$$

This value has been used to calculate the zooplankton biomass required to offset a given primary production (Fig. 7) or allochthonous input of particles (cf. STEEL) et al. 1972). The figure also shows the maximum biomass required, assuming that no zooplankton production occurs, and a minimum estimate using a somewhat unrealistic K_2 of 0.6. Zooplankton biomass calculated in this way correspond well with the values listed in Table 1. They are found in lakes, but only at certain times

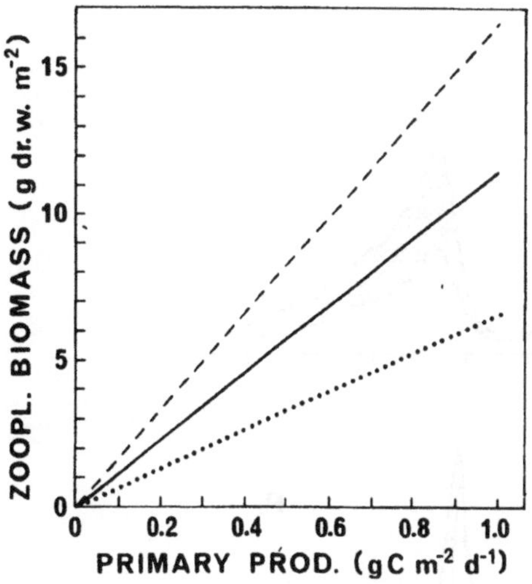

Fig. 7. The amount of zooplankton needed to balance the primary production as a function of primary productivity.

Estimate based on the metabolic requirements of zooplankton at 15 °C. The solid line represents the most reliable estimate using a K_2 of 0.3. The dashed line indicates the maximum estimate assuming no secondary production of the zooplankton. The dotted line is the result of the use of a K_2 of 0.6

of the year. Thus, the aim of the biomanipulation should be to expand this period and to maintain a sufficient density of zooplankton to offset the primary productivity of the lake.

Conclusions

There is a controversy whether it would be possible to increase the zooplankton standing crop during summer by reduction of fish. Certainly, zooplankton is not exclusively controlled by fish at this time of the year. During the clear-water phase, e.g., when zooplankton approaches the maximum biomass, egg numbers are very low, indicating that the quantity of food is limiting. Other mechanisms, like the interference of large inedible particles with the filtering apparatus of cladocerans (GLIWICZ & SIEDLAR 1980) or temperature-food interactions (THRELKELD 1986), may be important at times and contribute to the mid-summer-zooplankton depression (THRELKELD 1985). The success of a biomanipulation may depend largely on the relative importance of predation and food limitation for the zooplankton. However, during recent years several studies have shown that deliberate (BENNDORF et al. 1984, HENRIKSON et al. 1980, SHAPIRO & WRIGHT 1984) or accidental (DEBERNARDI & GIUSSANI 1977, DEBERNARDI 1981, SHAPIRO et al. 1982, LEAH et al. 1980) reductions of the fish stock led to a significant increase of the zooplankton standing crop, indicating that zooplankton density is partly determined by fish density, at least in eutrophic lakes.

Zooplankton can never eliminate all the particles in a lake, since it filters only nannoseston. If Fig. 7 is used to estimate the amount of zooplankton necessary to balance a certain primary production, this can only apply for the production of small size classes. Although often the majority of the carbon fixation takes place in particles smaller than 35 µm (RAI 1982), it has been reported that ungrazable algae and blue-greens can be favoured by heavy grazing stress (LYNCH 1980, HRBÁČEK 1964, MCCAULEY & BRIAND 1979, GANF 1983). The approach leading to Fig. 7 is extremely simple. There is

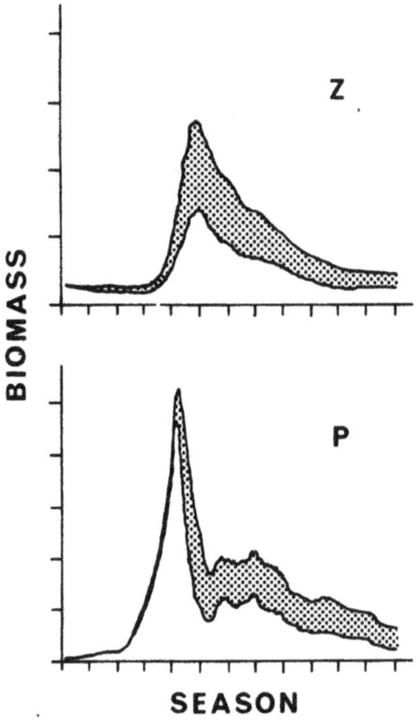

Fig. 8. An example of mathematical modelling of the effect of fish reduction on the seasonal abundance of phyto- and zooplankton biomass in a reservoir (redrawn from BENNDORF & RECKNAGEL 1979).

The shaded area indicates the difference in biomass resulting from a reduction of the fish stock by 80% (*Z* Zooplankton biomass, *P*. Phytoplankton biomass)

no experimental proof of this relationship. One has to consider that with respect to the aim of the manipulation (reduction of particles) increased grazing pressure causes positive (reduced photosynthetic biomass) and negative (nutrient recycling, greater light penetration, food limitation of grazers) feedbacks. Thus, these simple assumptions cannot predict the final equilibrium in a lake. Such a prediction should be made using the help of more complicated mathematical models (STRAŠKRABA 1979, BENNDORF & RECKNAGEL 1979, RECKNAGEL & BENNDORF 1982). A good example of what such predictions could look like is given in Fig. 8. This paper can only provide evidence that, even in eutrophic lakes, phytoplankton can be controlled by grazing and that the zooplankton densities required to do this are not unreasonably high.

Zusammenfassung

Das Zooplankton ist ein Schlüsselfaktor für die Biomanipulation in Seen. Die vorliegende Arbeit untersucht die Frage, wieviel Zooplankton für eine effektive Kontrolle des Sestons nötig ist. Die Beziehung zwischen zyklischen Transparenzphasen und der grazing-Aktivität wird am Schöhsee demonstriert. Es werden Angaben über quantitative Beziehungen zwischen Zooplanktonmasse und grazing-Effekt gemacht.

References

ANDERSON, C., BERGGREN, H., CRONBERG, G., & GELIN, C. (1978): Effects of planktivorous and benthivorous fish on organisms and water chemistry in eutrophic lakes. Hydrobiologia **59**, 9—15.

BAILEY-WATTS, A. (1982): The composition and abundance of phytoplankton in Loch Leven (Scotland) 1977—1979 and a comparison with the succession in earlier years. Int. Rev. ges. Hydrobiol. **67**, 1—25.

BARKO, J. W., BATES, D. J., FILBIN, G. J., HENNINGTON, S. M., & MCFARLAND, D. G. (1984): Seasonal growth and community composition of phytoplankton. J. Freshw. Ecol. **2**, 519—534.

BENNDORF, J., & RECKNAGEL, F. (1979): Experimente mit einem dynamischen ökologischen Modell der Freiwasserregion von Talsperren und Seen. Acta Hydrochim Hydrobiol. **7**, 473—490.

— KNESCHKE, H., KOSSATZ, K. & PENZ, E (1984): Manipulation of the pelagic food web by stocking with predacious fishes. Int. Rev. ges. Hydrobiol **69**, 407—428.

BLAAUBOER, M. C. I. (1982): The phytoplankton species composition and the seasonal periodicity in Lake Vechten from 1956 to 1979. Hydrobiologia **95**, 25—36.

BROOKS, J. L., & DODSON, S. I. (1965): Predation, body size, and composition of plankton. Science **150**, 28—35.

BÜRGI, H. R, WEBER, P., & BACHMANN, H., (1985): Seasonal variations in the trophic structure of phyto- and zooplankton in lakes in different trophic states. Schweiz. Z. Hydrol. **47**, 197 to 224.

CARPENTER, S., KITCHELL, J. F., & HODGSON, F. R. (1985): Cascading trophic interactions and lake productivity. BioScience **35**, 634—639.

CHOW-FRASER, P., & KNOECHEL, R., (1985): Factors regulating in situ filtering rates of Cladocera Can. J Fish. Aquat. Sci. **42**, 567—576.

CRUMPTON, W. G., & WETZEL, R. G. (1982): Effects of differential growth and mortality in the seasonal succession of phytoplankton population in Lawrence Lake, Michigan. Ecology **63**, 1729—1739.

DEBERNARDI, R. (1974): The dynamics of a population of *Daphnia hyalina* LEYDIG in Lago Maggiore, Northern Italy. Mem. Ist. Ital. Idrobiol. **31**, 221—243.

— (1981): Biotic interactions in freshwater and effects on community structure. Boll. Zool. **48**, 353—371.

— & GIUSSANI, G. (1977): The effect of mass fish mortality on zooplankton structure and dynamics in a small Italien lake (Lago di Annone). Verh. Int. Verein. Limnol. **20**, 1045—1048.

FOTT, J., PECHAR, L., & PRAŽÁKOVÁ, M. (1980a): Fish as a factor controlling water quality in ponds. Dev Hydrobiol. **2**, 255—261.

— DESORTOVÁ, B, & HRBÁČEK, J (1980b): A comparison of the growth of flagellates under heavy grazing stress with a continuous culture. In: SIKYTA, B., Y. FENCL & V. POLÁČEK (eds.), Continuous Cultivation of Microorganisms, Proc. 7th Symp. Prague, 395—401.

GANF, G. G. (1983): An ecological relationship between *Aphanizomenon* and *Daphnia pulex*. Austr. J. Mar. Freshw. Res. **34**, 755—773.

GELLER, W. (1980): Stabile Zeitmuster in der Planktonsukzession des Bodensees (Überlinger See). Verh. Ges. Ökol. **8**, 373—382.

GLIWICZ, Z. M. (1968): The use of anaestetizing substance in studies of food habits of zooplankton communities. Ekol. Pol. A **16**, 279—295.

— & HILLBRICHT-ILKOWSKA, A. (1975): Ecosystem of the Mikolajskie Lake. Elimination of phytoplankton biomass and its subsequent fate in lake through the year. Polskie Arch Hydrobiol. **22**, 39—52.

— & SIEDLAR, E. (1980): Food size limitation and algae interfering with food collection in *Daphnia*. Arch. Hydrobiol. **88**, 155—177

GREENE, C. H (1983): Selective predation in freshwater zooplankton communities Int. Rev. ges. Hydrobiol. **68**, 297—315.

GULATI, R. D. (1975): A study on the role of herbivorous zooplankton community as primary consumers of phytoplankton in Dutch lakes. Verh. Int Verein. Limnol. **19**, 1202—1210.

— (1983): Zooplankton and its grazing as indicators of trophic status in Dutch lakes. Env. Monit. Assessm. **3**, 343—354.

— (1984): The zooplankton and its grazing as measures of trophy in the Loosdrecht lakes. Verh. Int. Verein. Limnol. **22**, 863—867.

— SIEWERTSEN, K., & POSTEMA, G. (1982): The zooplankton, its community structure, food and feeding, and role in the ecosystem of Lake Vechten. Hydrobiologia **95**, 127—164.

— — — (1985): Zooplankton structure and grazing activities in relation to food quality and concentration in Dutch lakes. Arch. Hydrobiol., Beih. Ergebn. Limnol. **21**, 91—102.

HALL, D. J., THRELKELD, S. T., BURNS, C. W. & CROWLEY, P. H (1976): The size-efficiency hypothesis and the size structure of zooplankton communities. Ann. Rev. Ecol 7, 177—208

HANEY, J. F. (1973): An in situ examination of the grazing activities of natural zooplankton communities Arch Hydrobiol. **72**, 87—132

— & HALL, D. J. (1975). Diel vertical migration and filter-feeding activities of *Daphnia*. Arch. Hydrobiol. **75**, 413—441

HART, R. C. (1984): Zooplankton community grazing in silt-laden Lake Le Roux, Orange River, South Africa Verh. Int. Verein. Limnol. **22**, 1602—1607.

— (1986): Aspects of the feeding ecology of turbid water zooplankton. In situ studies of community filtration rates in silt-laden Lake le Roux, Orange River, South Africa. J. Plankton Res. **8**, 401—426.

HENRIKSON, L., NYMAN, H. G., OSCARSON, H. G., & STENSON, J. A. E. (1980): Trophic changes without changes in the external nutrient loading. Hydrobiologia **68**, 257—263.

HORN, W. (1981): Phytoplankton losses due to zooplankton grazing in a drinking water reservoir. Int. Rev. ges. Hydrobiol. **66**, 787—810.

HRBÁČEK, J (1962): Species composition and the amount of the zooplankton in relation to the fish stock. Rozpr. ČSAV, řada matem. a přir. věd. **72**, 1—116.

— (1964): Contribution to the ecology of water-bloom-forming blue-green algae *Aphanizomenon flos-aquae* and *Microcystis aeruginosa*. Verh. Int. Verein. Limnol. **15**, 837—846.

KNISELY, K., & GELLER, W. (1986): Selective feeding of four zooplankton species on natural lake phytoplankton. Oecologia (Berlin) **69**, 86—94.

KNOECHEL, R., & HOLTBY, L. B. (1986): Construction and validation of a body-length-based model for the prediction of cladoceran community filtering rates. Limnol. Oceanogr. **31**, 1—16.

KRAMBECK, H. J., HICKEL, B., HOFMANN, W., & OVERBECK, J. (1978): Mathematische Modelle als integrierende Hilfsmittel des Limnologen, dargestellt am Ökosystem Plußsee. Verh. Ges. Ökol. **7**, 137—144.

LAMPERT, W. (1977) Studies on the carbon balance of *Daphnia pulex* DE GEER. II. The dependence of carbon assimilation on animal size, temperature, food concentration and diet species Arch. Hydrobiol., Suppl. **48**, 310—335.

— (1978): Climatic conditions and planktonic interaction as factors controlling the regular succession of spring algal bloom and extremely clear water in Lake Constance. Verh. Int. Verein. Limnol. **20**, 969—974.

— (1984): The measurement of respiration. In: DOWNING, J. A., & RIGLER, F. H (eds), A Manual on Methods for the Assessment of Secondary Production in Fresh Water, IBP Handbook No 17, 2nd edn. Oxford, pp. 413—468.

— & TAYLOR, B. E (1984): In situ grazing rates and particle selection by zooplankton: Effects of vertical migration. Verh. Int Verein. Limnol **22**, 943—946.

— — (1985): Zooplankton grazing in a eutrophic lake: implications of vertical migration. Ecology **66**, 68—82.

— FLECKNER, W., RAI, H., & TAYLOR, B. E. (1986): Phytoplankton control by grazing zooplankton: a study on the clear-water phase. Limnol Oceanogr. **31**, 478—490.

LEAH, R. T., MOSS, B., & FORREST, D. E. (1980): The role of predation in causing major changes in the limnology of a hyper-eutrophic lake. Int. Rev. ges. Hydrobiol **65**, 223—247

LEHMAN, J. T. (1980): Release and cycling of nutrients between planktonic algae and herbivores. Limnol Oceanogr. **25**, 620—632.

LYNCH, M. (1980) *Aphanizomenon* blooms Alternate control and cultivation by *Daphnia pulex*. In: KERFOOT, W. C. (ed.), Evolution and Ecology of Zooplankton Communities. Hanover, N.H., pp. 299—304.

MCCAULEY, E., & BRIAND, F. (1979): Zooplankton grazing and phytoplankton species richness: Field test of the predation hypothesis. Limnol. Oceanogr. **24**, 243—252.

— & KALFF, J. (1981): Empirical relationship between phytoplankton and zooplankton biomass in lakes. Can. J. Fish Aquat. Sci. **38**, 458—462.

MCQUEEN, D J., POST, J R., & MILLS, E L (1986) Trophic relationships in freshwater pelagic ecosystems Can. J. Fish Aquat Sci. **43**, 1571—1581.

MEFFERT, M. E., & OVERBECK, J (1985): Dynamics of chlorophyll and photosynthesis in natural phytoplankton associations I Seasonal and annual cycles of chlorophyll and phaeopigments, in relation to euphotic depth and their vertical distribution in small Northgerman lakes. Arch. Hydrobiol. **104**, 219—234.

NOWAK, K.-E. (1975). Die Bedeutung des Zooplanktons für den Stoffhaushalt des Schierensees. Arch. Hydrobiol **75**, 149—224.

OKAMOTO, K (1984). Size-selective feeding of *Daphnia longispina* and *Eudiaptomus japonicus* on a natural phytoplankton assemblage with the fractionizing method Mem. Fac. Sci. Kyoto Univ., Ser Biol. **9**, 23—40.

OSKAM, G. (1978) Die Vorausberechnung der Algenbiomasse in den Biesbosch-Speicherbecken. Theorie und Praxis. In: OSKAM, G., & BERNHARDT, H (Hrsg.), Verminderung der Algenentwicklung in Speicherbecken und Talsperren. DVGW-Schriftenreihe Wasser Nr. 16, 90—107.

PETERS, R. H., & DOWNING, J. A. (1984): Empirical analysis of zooplankton filtering and feeding rates Limnol Oceanogr. **29**, 763—784.

PETERSEN, F (1983): Population dynamics and production of *Daphnia galeata* (Crustacea, Cladocera) in Lake Esrom. Holarct. Ecol. **6**, 285—294.

PORTER, K. G. (1973): Selective grazing and differential digestion of algae by zooplankton. Nature **244**, 179—180.

RAI, H. (1982): Primary production of various size fractions of natural phytoplankton communities in a North German lake. Arch. Hydrobiol. **95**, 395—412.

RECKNAGEL, F., & BENNDORF, J. (1982) Validation of the ecological simulation model SALMO. Int. Rev. ges. Hydrobiol. **67**, 113—125

REINERTSEN, H., & LANGELAND, A. (1982): The effect of lake fertilization on the stability and material utilization of a limnetic ecosystem. Holarct Ecol. **5**, 311—324

— & OLSEN, Y. (1984): Effects of fish elimination on the phytoplankton community of a eutrophic lake. Verh. Int. Verein. Limnol **22**, 649—657.

REYNOLDS, C. S. (1984): The ecology of freshwater phytoplankton. Cambridge, 384 pp.

— THOMPSON, J. M., FERGUSON, A. J. D., & WISEMAN, S. W. (1982): Loss processes in the dynamics of phytoplankton maintained in closed systems. J. Plankton Res. **4**, 561—600.

RIEMANN, B. (1983): Biomass and production of phyto- and bacterioplankton in eutrophic Lake Tystrup, Denmark. Freshw. Biol. **13**, 389—398

ROSS, P. E., & MUNAWAR, M (1981): Preference for nannoplankton size fractions in Lake Ontario zooplankton grazing. J. Great Lakes Res **7**, 65—67

SCHOBER, U. (1980): Kausalanalytische Untersuchungen der Abundanzschwankungen des Crustaceen-Planktons im Bodensee. Diss., Univ. Freiburg/Brsg., 162 pp.

SHAPIRO, J. (1980): The importance of trophic-level interactions to the abundance and species composition of algae in lakes Dev Hydrobiol. **2**, 105—116.

— & WRIGHT, D. I. (1984): Lake restoration by biomanipulation: Round Lake, Minnesota, the first two years Freshw Biol **14**, 371—383.

— FORSBERG, B., LAMARRA, V., LINDMARK, G., LYNCH, M., SMELTZER, E., & ZOTO, G (1982): Experiments and experiences in biomanipulation. Interim Report of the Limnological Research Center, University of Minnesota, Minneapolis **19**.

SOMMER, U (1984): Light, stratification and zooplankton as controlling factors for the spring development of phytoplankton in Lake Constance Schweiz. Z. Hydrol. **45**, 394—404.

— GLIWICZ, Z. M., & LAMPERT, W. (1985). The PEG model of seasonal succession of plankton in fresh waters. Arch. Hydrobiol. **106**, 433—471.

STEEL, J. A. P., DUNCAN, A., & ANDREW, T. E. (1972). The daily carbon gains and losses in the seston of Queen Mary Reservoir, England, during early and mid 1968. In: KAJAK, Z., & A. HILLBRICHT-ILKOWSKA (eds.), Productivity Problems of Freshwaters, PWN-Polish Scientific Publ., Warszawa-Krakow, pp. 515—527.

STERNER, R. W. (1986). Herbivores' direct and indirect effects on algal populations. Science **231**, 605—607.

STRAŠKRABA, M. (1979): Mathematische Simulation der Produktionsdynamik in Gewässern und deren Anwendung auf die Produktionssteuerung in Talsperren. Z. Wasser-Abwasser-forsch. 12, 56—64.

THOMPSON, J. M., FERGUSON, A. J. D., & REYNOLDS, C. S. (1982): Natural filtration rates of zooplankton in a closed system: the derivation of a community grazing index. J. Plankton Res. 4, 545—560.

THRELKELD, S. T. (1985): Resource variation and the initiation of midsummer declines of cladoceran populations. Arch. Hydrobiol., Beih. Ergebn. Limnol. 21, 333—340.

— (1986): Differential temperature sensitivity of two cladoceran species to resource variation during a blue-green algal bloom. Can J Zool 64, 1739—1744.

— (1987): Experimental evaluation of trophic-cascade and nutrient mediated effects of planktivorous fish on plankton community structure. In: KERFOOT, W C, & A. SIH (eds.), Predation. Direct and indirect impacts on aquatic communities. Hanover, N.H, pp. 161—173.

TILZER, M M. (1984): Estimation of phytoplankton loss rates from daily photosynthetic rates and observed biomass changes in Lake Constance. J. Plankton Res. 6, 309—324.

TSCHUMI, P-A, BANGERTER, B. & ZBÄREN, D. (1982): Zehn Jahre limnologische Forschung am Bielersee (1972—1981). Vierteljahresschr. Naturforsch. Ges. Zürich 127, 337—355.

UHLMANN, D. (1971): Influence of dilution, sinking and grazing rate on phytoplankton populations of hyperfertilized ponds and microecosystems. Mitt. Int. Verein. Limnol. 19, 100—124.

VANDERPLOEG, H. A., SCAVIA, D., & LIEBIG, J. R. (1984): Feeding rate of *Diaptomus sicilis* and its relation to selectivity and effective food concentration in algal mixtures and in Lake Michigan. J. Plankton Res. 6, 943—952.

WINBERG, G. G (ed.), (1971): Methods for the estimation of production of aquatic animals. London, New York, 175 pp.

WRIGHT, J. C. (1958): The Limnology of Canyon Ferry Reservoir. I. Phytoplankton-zooplankton relationships in the euphotic zone during September and October 1956. Limnol. Oceanogr. 3, 150—159.

ZARET, T M. (1980): Predation and freshwater communities New Haven, 187 pp.

Received January 9, 1987.

Address of the author:
Prof. Dr. W. LAMPERT
Abt. Ökophysiologie
Max-Planck-Institut für Limnologie
Postfach 165
D-2320 Plön

Limnologica (Berlin) 19 (1), 21—26, Juli 1988

Department of Hydrobiology, University of Warsaw, Warsaw, Poland; "Vijverhof" Laboratory, Limnological Institute, Nieuwersluis, The Netherlands

Can *Daphnia* Prevent a Blue-green Algal Bloom in Hypertrophic Lakes? A Laboratory Test

By Piotr Davidowicz, Z. Maciej Gliwicz & Ramesh D. Gulati

With 7 Figures

Key words: *Daphnia magna*; Filamentous blue-green algae; Hypertrophic lakes; Biomanipulation; Filtering rate; Fecundity

Abstract

Daphnia magna was exposed to various concentrations of filamentous blue-green algae in continous flow system of interconnected aquaria. *Daphnia* caused a significant decrease in filament concentration and influenced the length distribution of filaments, by breaking up the longest trichomes. The effect of *Daphnia* of filaments was increased with decreased algal abundance. Increase in concentration of blue-green algae caused decrease in filtering rate and fecundity of *Daphnia* — when algal density was high (150 µg chlorophyll a per liter) reproduction was halted. Thus, the high concentration of filamentous blue-green algae should be considered as strong factor influencing *Daphnia* demography.

Introduction

Phytoplankton blooms in eutrophic lakes are most commonly composed of algal species that can not be efficiently controlled by filter-feeding zooplankton. This is because such algae have a form which is difficult to handle, have large cell or colony size, or are toxic to animals (Gliwicz 1977, Lampert 1977, Briand & McCauley 1978, Porter & Orcutt 1980, Lampert 1981a, 1981b). Bloom-forming filamentous blue-green algae are not only unmanageable as food for zooplankton, but may also interfere with the process of food collection in cladoceran species (Gliwicz 1977, Webster & Peters 1978).

Some workers belive that algal populations in eutrophic lakes cannot be controlled by zooplankton, because large filter-feeding cladoceran species, e.g. *Daphnia magna*, *D. pulex*, or *D. pulicaria* are never abundant, being under pressure exerted by planktivorous fish selecting large prey (Hrbáček et al. 1961, Shapiro et al. 1975, Lynch & Shapiro 1981). As a consequence large cladocerans contribute only marginally to zooplankton community grazing in highly eutrophic lakes (Gulati et al. 1985). These large cladoceran species, because of their high filtering rate and wide spectrum of ingested particles (Burns 1969, Geller & Müller 1981) should be able to control large and filamentous algae to greater extent than small cladocerans and rotifers, which are common in eutrophic lakes.

The aim of this paper is to test whether large *Daphnia* can control density of algal filaments and to find the maximum concentration of filamentous blue-green algae at which *Daphnia magna* is able to survive and increase its population density.

Material and methods

Lake water from highly eutrophic Loosdrecht Lakes (The Netherlands), inhabited by dense — up to 300 µg chlorophyll a per liter — populations of *Oscillatoria limnetica*, *O. agardhii*, *O. redekei*, *Aphanizomenon flos-aquae* and *Lyngbya* sp. (Gulati et al. 1985) was used in the experiments. It was prefiltered through 200 µm net to remove native crustacean herbivore zooplankton and than continously pumped into two sets of interconnected 5-l aquaria, filled initially with water devoid of algae. *Daphnia magna* from clonal laboratory culture was introduced to one set, while the second set served as a control. The outflows from aquaria were covered with 400 µm net to prevent *Daphnia* washout. Two experiments were performed. They differed in the number of aquaria in each set (6 and 5), flow rate (5 and 2.5 l per day) and in the density of *Daphnia* inocula (15 and 30 ind per liter). The water temperature (20 °C), light conditions (13 W · m^2) and light: dark period (12:12 h) were similar in both the experiments.

Chlorophyll a concentration in water flowing into the system and leaving the last aquarium was monitored daily with Zeiss spectrophotometer after ethanol extraction. Spectrophotometric measurements served as calibration for fluorometric evaluations of chlorophyll a concentration in each aquarium of the two sets, performed in vivo with Aminco-Bowman Spectrofluorometer.

Simultaneously the number and the length of algal filaments were determined in the first and the last aquaria from both sets under Zeiss inverted microscope. Highly significant correlation between chlorophyll a concentration and filaments density (r, 0.93; p, 0.001), expressed by the regression equation:

$$d = 516c + 5$$

where d is filaments density in ml^{-1} and c is chlorophyll a concentration in µg · l^{-1}, allowed to estimate the number of filaments in intermediate aquaria.

Daphnia density in each experimental aquarium was evaluated daily. The walls of aquaria — the one in front and one opposing it — were divided into 16 squares, and individuals of *Daphnia* found within the corresponding squares were counted visually. The sum of 16 squares gave total number of *Daphnia* in aquarium.

Gut contents of *Daphnia* from each aquarium were also analysed, and the length of algal filaments found in guts and in aquarium water were measured under Zeiss inverted microscope. In the second experiment filtering rates of *Daphnia* exposed to water from subsequent aquaria were measured at the end, employing the ^{14}C-tracer technique (Gulati et al. 1982). The seston fraction <33 µm was labelled with $NaH^{14}CO_3$ for 3 days.

Results

The concentration of algal filaments in each aquarium of both the experimental and control sets gradually increased (Fig. 1). Compared with the control, the increase in experimental aquaria was significantly lower in last two aquaria of the first experiment ($p = 0.01$ and 0.005 for E5 and E6, respectively) and the last four aquaria of the second experiment ($p = 0.05, 0.01, 0.005$ and 0.001 for E2, E3, E4 and E5, respectively) (Mann-Whitney *U*-test; Fig. 2). This could be caused by *Daphnia* grazing.

Daphnia ingested filaments of all the blue-green species present in the medium and selected the shortest ones.

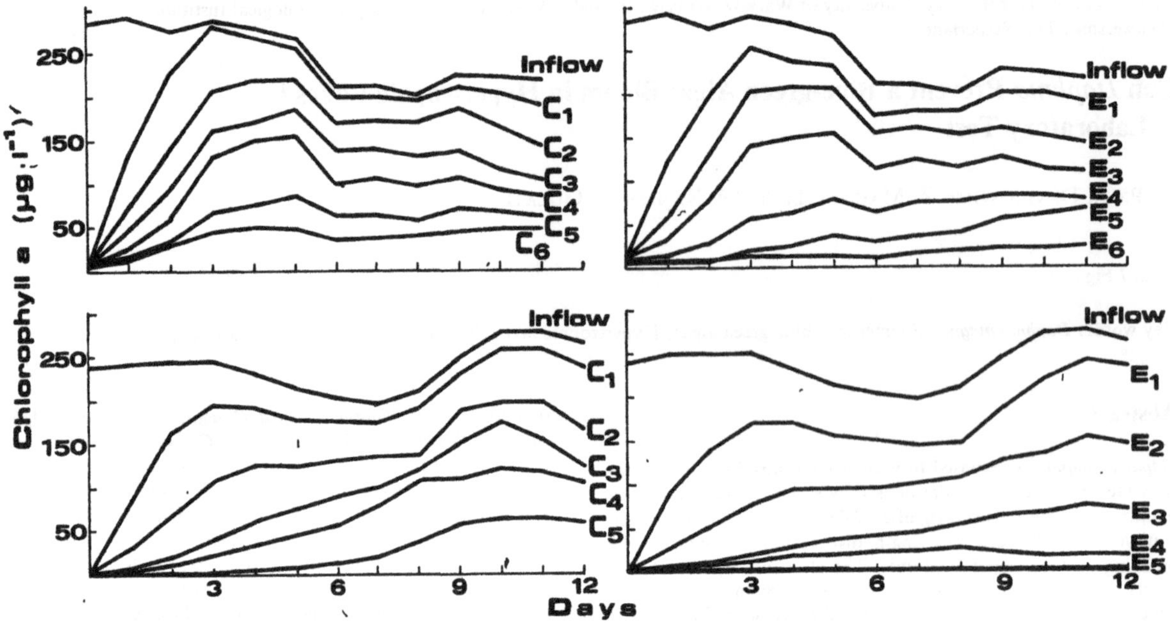

Fig. 1. Changes in concentration of filaments, expressed as chlorophyll a concentration, in the inflow and each subsequent aquarium of the control (left) and of the experimental, i e. with *Daphnia* added (right). First experiment — top panels; second experiment — bottom panels.

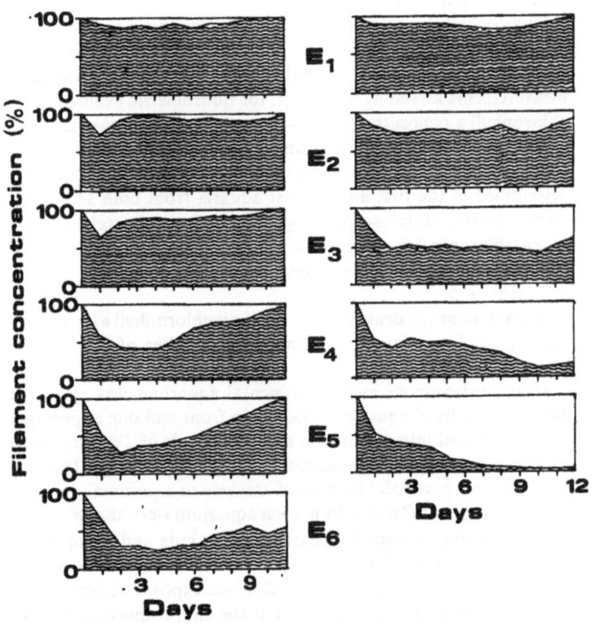

Fig 2 Changes in the filament concentration in different aquaria of the experimental set, expressed as % of the chlorophyll a concentration in the corresponding control aquaria (e g. the concentration in El is expressed as % of concentration in Cl) Left panels — first experiment, right panels — second experiment.

This can be judged from the comparison of the filament length distribution in the medium and in *Daphnia* intestines (Fig. 3). Due to *Daphnia* filtering activity the mean length of filaments decreased in subsequent aquaria. This made the filament suspension easier to handle for *Daphnia* in aquaria at the end of the set (for details see DAWIDOWICZ, submitted).

The decrease in both filament concentration and in mean filament length led to an increased elimination of blue-green algae by *Daphnia* in subsequent aquaria. This effect diminished in course of time in all aquaria in the first experiment. In the first three aquaria (E 1, E2, E3) it became negligible after the first two days (Fig. 2, left) when filament concentration increased to about 200 µm chlorophyll a per liter (Fig. 1, top). The effect was greater in second experiment (Fig. 2, right) in which the increase in the filament concentration was more gradual (Fig. 1, bottom) due to reduced flow and higher initial density of *Daphnia*. In the last two aquaria (E4 and E5) the effect became even stronger at the end of the experiment, which suggests the possibility of continous and increased control of filaments by *Daphnia*. The concentrations of blue-green filaments in these two aquaria never exceed 20 µm chlorophyll a per liter (Fig. 1, bottom right) and *Daphnia* density was increasing throughout the experiment (Fig. 4, right). Some decrease in *Daphnia* density was already detected in the preceding aquarium (E3) and a constant decrease was seen in the first two quaria of the set (E 1 and E2). Such differences were observed also in first experiment (Fig. 4, left).

In both experiments *Daphnia* density was reduced nearly to zero in the first aquarium of each of the two experimental sets (Fig. 4). At the end of experiment there were no juvenile animals left, nor eggs were present in adults brood chambers (Fig. 5). In the second aquarium in each of the experiments both the share of juveniles and the fecundity also significantly decreased (E2, Figs. 4 and 5).

The fecundity of *Daphnia* appeared to be affected by much lower concentrations of filaments than those observed in the first two aquaria. This became evident from the regressions between the clutch size and the filament concentration for the two experiments ($r = -0.997$ and $r = -0.975$, $p = 0.001$ in both cases; Fig. 5).

The filtering rates of all four size classes of *Daphnia* were strongly reduced as the concentration of filaments increased (Fig. 6). Smaller individuals were relatively more sensitive to increased filament density. Their filtering rate was reduced to one-fourth with an increase in filament concentration from 3 to 18 µg of chlorophyll a per liter (Fig. 7).

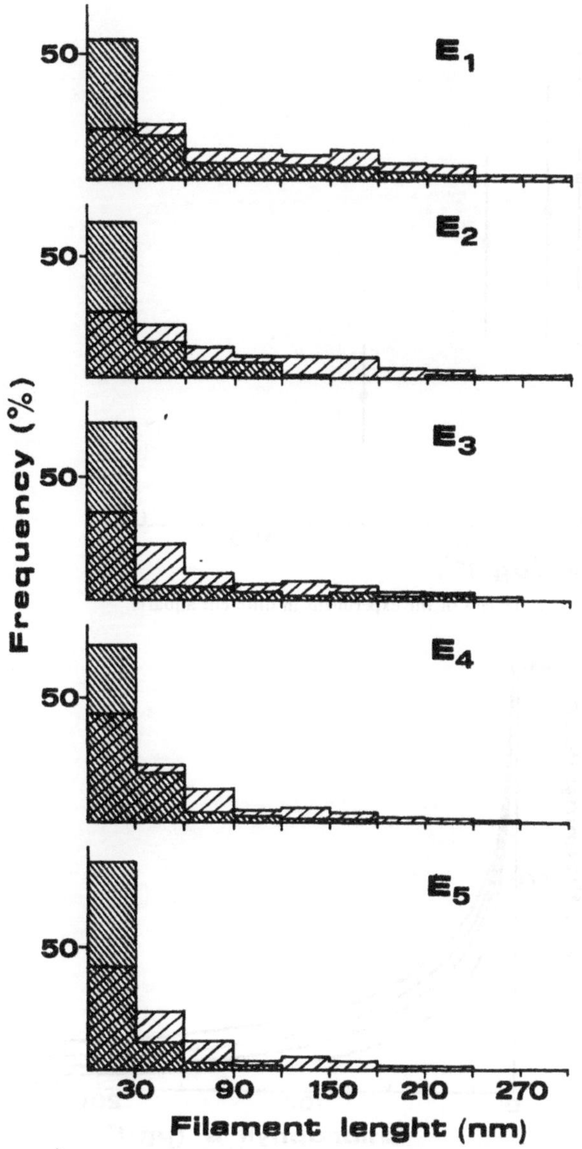

Fig. 3. Frequency distribution of filament lengths in the medium (right shaded) and in *Daphnia* intestines (dark shaded) in subsequent aquaria (E1—E5) of the experimental set. Data collected at the end of second experiment.

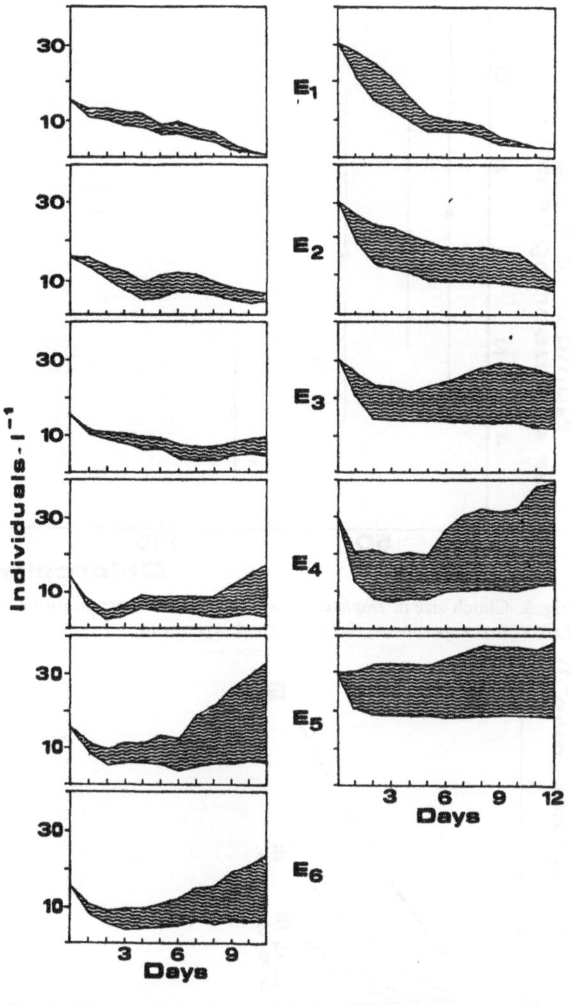

Fig. 4. Changes in *Daphnia* density in different aquaria of the experimental set in the first experiment (left panel) and in the second experiment (right panel). Share of juveniles (below 1.5 mm body length) indicated by the striped area

Discussion

Strong negative effects of high concentrations of filamentous blue-green algae on cladocerans are well known. Blue-green algal filaments, even if not toxic, have low nutritional value (LAMPERT 1977, BRIAND & MCCAULEY 1978, HOLM et al. 1983), and may interfere with filtering mechanisms reducing food intake of large cladocerans (GLIWICZ 1980, GLIWICZ & SIEDLAR 1980). Increased energy expenditure for rejection of filaments from filter appendages and food groove may also reduce net assimilation rate, growth and reproduction (PORTER et al. 1983, PORTER & MCDONOUGH 1984). The inhibitory effect of filamentous blue-greens has been suggested as an explanation for the absence of large, most sensitive species of *Daphnia* from eutrophic lakes (EDMONDSON & LITT 1982, INFANTE & ABELLA 1985, GLIWICZ 1985). This explanation, alternative to the more frequently cited explanation of size selective predation (BROOKS & DODSON 1965, WERNER & HALL 1974) by planktivorous fishes, is confirmed by our data.

The experimental populations of *Daphnia magna* disappeared after two weeks of exposure to water containing blue-green algae in concentrations similar to those observed in Loosdrecht Lakes (first aquarium in both experiments). Fecundity of *Daphnia* decreased sharply with increasing algal density and at concentration of blue-greens corresponding to 150 µg chlorophyll a per liter the animals stopped reproducing. A two fold decrease in algal concentration allowed *Daphnia* to survive, but the observed fecundity was very low, clutch size being about 1 egg per female, the density of population did not exceed the initial level.

The *Daphnia* disappearance from the first aquarium in both experiments, which had relatively the highest filament concentration, could not be attributed to the effect of toxins excreted by blue-green algae to the medium. *Daphnia* populations in the last aquaria, which contained the same water as the first one but much less filaments, survived well and increased their density.

Therefore, high density of filamentous blue-green algae seem to strongly affect *Daphnia* demography, by limiting fecundity and reproduction rate. This observation may have important consequences for applying the

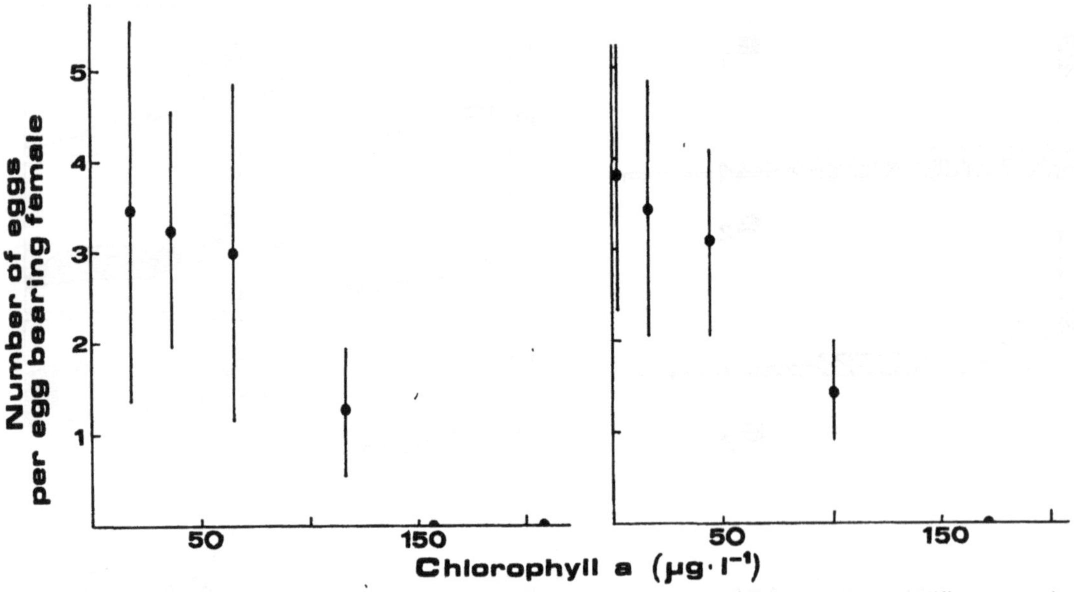

Fig. 5. Clutch size in *Daphnia* at various filament concentrations observed at the end of the experiment in different aquaria. Left — first experiment; right — second experiment.

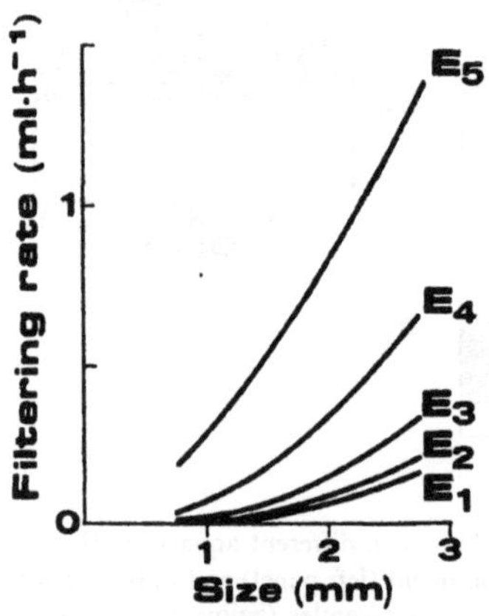

Fig. 6. Changes in the filtering rates of *Daphnia* with increase in body length at various filament concentrations present in the aquaria of the experimental set at the end of the second experiment. The food concentrations were 221, 137, 69, 18, and 3 μg chlorophyll a per liter in aquaria E1, E2, E3, E4, and E5, respectively. All regressions significant at $p = 0.001$ ($y = 0.01x^{2.71}$ for E1, $y = 0.011x^{2.90}$ for E2, $y = 0.033x^{2.29}$ for E3, $y = 0.083x^{2.05}$ for E4, and $y = 0.299x^{1.51}$ for E5); x and y are length and filtering rate of *Daphnia*, respectively.

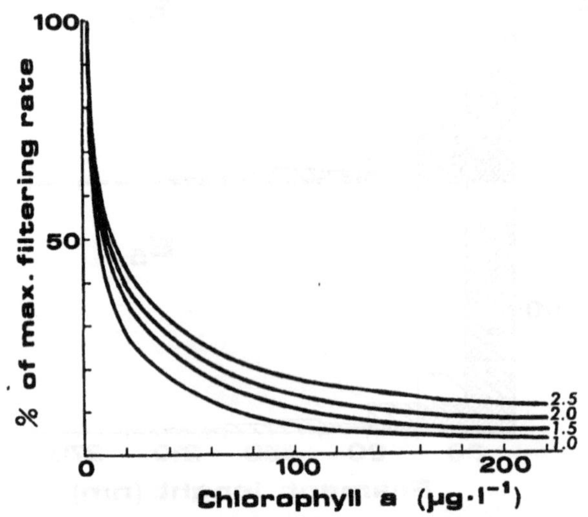

Fig. 7. Changes in the filtering rates of four size classes of *Daphnia* (2.5, 2.0, 1.5, and 1.0 mm body length) with the increase in filament concentration. Filtering rate is expressed as % of the rate observed for each size classes at the lowest filament concentration (3 μg chlorophyll a perliter).

biomanipulation methods to hypertrophic lakes for their recovery.

The essence of biomanipulation approach is to reduce planktivorous fish pressure to allow an increased growth in populations of large planktonic filter-feeders. Large cladocerans, due to their high filtering rate and wide spectrum of particles they can ingest, should keep the density of phytoplankton at low level, even in nutrient rich environments (SHAPIRO et al. 1975). This control of algal abundance, though not efficient in case of some totally ungrazeable algae, like flake form of *Aphanizomenon* (LYNCH 1980), is usually strong enough to support long clear water phase (FOTT et al. 1980, SHAPIRO et al. 1982).

However, in lakes inhabited by blue-green algae in concentrations as high as in the Loosdrecht Lakes, even total elimination of fish pressure may not cause any immediate shift in zooplankton towards domination of large cladocerans. At the concentrations of filaments occuring in these lakes adult *Daphnia magna* can survice for a few days only, juvenile stages being more susceptible. Even if *Daphnia* were to persist for longer time in these conditions, it will not be able to influence algal abundance markedly, because its filtering rate is strongly suppressed by blue-green filaments.

On the other hand *Daphnia* can ingest blue-green algal filaments and (as suggested elsewhere — DAWIDOWICZ, submitted) is able to break up long trichomes, making them more available for filter-feeding zooplankton. This may allow a control of algal populations if the initial density of filaments is not so high that it reduces strongly both filtering rate and fecundity of *Daphnia*. In other words the efficiency of algal density control decreases rapidly with increasing abundance of filaments. In our study an increase in algal concentration, chlorophyll increasing from 5 to 20 μg per liter, caused more than two-fold decrease in the filtering rate. A further increase in the filament abundance does not suppress *Daphnia* filtering rate so strongly, but diminishes its fecundity drastically, thus reducing the possibility of demographic response of *Daphnia* population to changing algal concentration.

The Loosdrecht Lakes are extremely eutrophic with dense blue-green blooms in summer. It seems, that even in such hypertrophic environment algal concentration could be kept at a low level. This could be accomplished if a large-sized *Daphnia* population became abundant and persist in the lake since early spring, before the bloom starts building up. It is possible however only if fish will not wipe the *Daphnia* population out in the summer period.

Summary

The aim of the present paper is to test whether large *Daphnia* is able to control the abundance of blue-green algal filaments and to find the maximum concentration of these algae at which *Daphnia magna* can survive and increase its population density.

Water from hypertrophic Loosdrecht Lakes, dominated by blue-green algae, was passed through two sets of interconnected aquaria. *Daphnia magna* was introduced to one set, with a second set serving as a control Number of *Daphnia* in experimental aquaria as well as algal density in both sets were monitored daily. After a week of exposure the gut contents of *Daphnia* from different aquaria were determined and compared with composition of the seston in the aquaria. Filtering rates of four size classes of *Daphnia* exposed to varying concentrations of blue-green filaments were measured, employing ^{14}C-tracer technique.

The density of filaments in the experimental set was significantly lower than in control, certainly due to *Daphnia* grazing. *Daphnia* ingested filaments of all blue-green algal species present in the medium and selected shortest ones (<33 μm). The effect of *Daphnia* on filaments density was increased with decreased algal abundance in the subsequent aquaria. *Daphnia* density was markedly reduced in the first aquarium of the set, in which a concentration of blue-green algae was very high, corresponding to 200 μg chlorophyll a per liter, but continuously increased up to the last aquarium, which had algal concentrations only one-tenth of those in the first one. Both, fecundity and filtering rate of *Daphnia* were strongly suppressed by increasing filaments abundance.

It seems, that *Daphnia magna* can control population growth of filamentous blue-green algae in nutrient-rich lake water. It can however do so only, if its population density is high enough to prevent blue-greens to reach the concentration at which filtering rate and food ingestion of *Daphnia* are reduced below the threshold level necessary for the minimum reproduction required for maintaining its dense population

When a natural community of filamentous blue-green algae approaches a concentration corresponding to about 150 μg chlorophyll a per liter the reproduction of *Daphnia* ceases and its population disappears from the environments in few days.

Zusammenfassung

In einem Labor-Durchflußsystem werden die Beziehungen von *Daphnia magna* zu fädigen Blaualgen untersucht. *Daphnia* bewirkt eine Abnahme der Algendichte und durch Abbrechen von Trichomen eine Änderung der Längenverteilung der Fäden. Hohe Algendichten führen jedoch zu einer Reduzierung der Filterrate und der Reproduktion der Daphnien.

Acknowledgement

The first two authors feel obliged to the Limnological Institute, Nieuwersluis, for providing travel founds and to colleagues at the Institute who helped in various ways.

References

Briand, F., & McCauley, E. (1978): Cybernetic mechanisms in lake plankton system: how to control undesirable algae. Nature **273**, 228—230.

Brooks, J. L., & Dodson, S. I. (1965): Predation, body size and composition of plankton. Science **150**, 28—35.

Burns, C. W. (1969): Relation between filtering rate, temperature and body size in four species of *Daphnia*. Limnol. Oceanogr. **14**, 693—700

Edmondson, W. T., & Litt, A. H. (1982): *Daphnia* in Lake Washington. Limnol Oceanogr. **27**, 272—293

Fott, J, Pechar, J., & Prazakova, M. (1980)· Fish as factor controlling water quality in ponds Dev. Hydrobiol. **2**, 255—261.

Geller, W, & Müller, H (1981): The filtration apparatus of Cladocera filter mesh sizes and their implications on food selectivity. Oecologia (Berlin) **49**, 316—321

Gliwicz, Z. M. (1977): Food size selection and seasonal succession of filter feeding zooplankton in an eutrophic lake. Ekol. Pol. **25**, 179—225.

— (1980): Filtering rates, food size selection, and feeding rates in cladocerans — another aspect of interspecific competition in filter feeding zooplankton. Am. Soc. Limnol. Oceanogr. Spec. Symp **3**, 282—291.

— (1985): Predation or food limitation: an ultimate reason for extinction of planktonic cladoceran species. Arch. Hydrobiol, Beih Ergeb. Limnol. **21**, 419—430.

— & Siedlar, E. (1980): Food size limitation and algae interfering with food collection in *Daphnia* Arch Hydrobiol. **88**, 155—177.

Gulati, R. D., Siewertsen, K., & Postema, G. (1982): The zooplankton: Its community structure, food and feeding and role in the ecosystem of lake Vechten. Hydrobiologia **95**, 127—163

— — — (1985): Zooplankton structure and grazing activities in relation to food quality and concentration in Dutch Lakes. Arch. Hydrobiol., Beih. Ergebn. Limnol. **21**, 91—102

Holm, N. P., Ganf, G. G., & Shapiro, J. (1983): Feeding and assimilation rates of *Daphnia pulex* fed *Aphanizomenon flos-aquae*. Limnol. Oceanogr. **28**, 677—687.

Hrbáček, J., Dvorakova, M., Korinek, V., & Prohazkova, L. (1961): Demonstration of the effect of the fish stock on the species composition and the intensity of metabolism of the whole plankton association. Verh. Int. Verein. Limnol. **143**, 192—195.

Infante, A., & Abella, S. E. B. (1985): Inhibition of *Daphnia* by *Oscillatoria* in Lake Washington. Limnol. Oceanogr. **30**, 1046—1052.

Lampert, W. (1977): Studies on the carbon balance of *Daphnia pulex* De Geer as related to environmental conditions. II. The dependance of carbon assimilation on animal size, temperature, food concentration and diet species. Arch Hydrobiol. Suppl. **48**, 310—335.

— (1981a): Inhibitory and toxic effect of blue-green algae on *Daphnia*. Int. Rev. ges. Hydrobiol **66**, 285—298.

— (1981b)· Toxicity of the blue-green *Microcystis aeruginosa*: effective defence mechanism against grazing pressure by *Daphnia*. Verh. Int. Verein. Theor. Angew. Limnol. **21**, 1436 to 1440.

Lynch, M. (1980)· *Aphanizomenon* blooms: alternate control and cultivation by *Daphnia pulex*. Am. Soc. Limnol. Oceanogr. Spec. Symp. **3**, 299—304.

— & Shapiro, J. (1981)· Predation, enrichment, and phytoplankton community structure: an experimental study. Limnol. Oceanogr. **26**, 86—102.

Porter, K. G., & McDonough, R. (1984): The energetic cost of response to blue-green algal filaments by cladocerans. Limnol. Oceanogr. **29**, 365—369.

— & Orcutt, J. D. (1980): Nutritional adequacy, manageability, and toxicity as factors that determine the food quality of green and blue-green algae for *Daphnia*. Am. Soc. Limnol. Oceanogr. Spec. Symp. **3**, 268—281.

— — & J. Gerritsen (1983): Functional response and fitness in a general filter feeder, *Daphnia magna* (Cladocera: Crustacea). Ecology **64**, 735—742.

Shapiro, J., Forsberg, B., Lamarra, V., Lindmark, G., Lynch, M., Smeltzer, E., & Zoto, G. (1982): Experiments and ex-

periences in biomanipulation · Studies of ways to reduce algal abundance and eliminate bluegreens. Interim Rep. No. 19, Limnol. Res. Center Univ. of Minnesota, Minneapolis, 251 pp.
Shapiro, J, Lamarra, V., & Lynch, M., (1975): Biomanipulation: An ecosystem approach to lake restauration. In: Brezonik, P. L., & Fox, J. L. (eds.), Proc. symp. on water quality management through biological control. Rep. No. ENV-07-75-1, University of Florida, Gainsville, 85—96.
Webster, K. E., & Peters, R. H (1978): Some size-dependent inhibition of larger cladoceran filterers in filamentous suspensions. Limnol. Oceanogr. 23, 1238—1245.
Werner, E. E., & Hall, D. J. (1974): Optimal foraging and the size selection of prey by the bluegill sunfish (*Lepomis macrochirus*). Ecology 55, 1042—1052.

Received November 28, 1986.

Address of the authors:

P. Dawidowicz
Z. M Gliwicz
Department of Hydrobiology
University of Warsaw
Nowy Swiat 67
00-046 Warsaw
Poland

R. D. Gulati
"Vijverhof" Laboratory
Limnological Institute
Rijksstraatweg 6
3631 AC Nieuwersluis
The Netherlands

Limnologica (Berlin) 19 (1), 27—35, Juli 1988

Institute of Limnology, Uppsala, Sweden

Restoration of the Pelagic Food Web in Acidified and Limed Lakes by Gentle Fertilization

By H. OLOFSSON, P. BLOMQVIST, H. OLSSON & O. BROBERG

With 2 Figures and 6 Tables

Key words · Planktonic structure; Acidified lakes; Liming, Fertilization

Abstract

Liming and subsequent fertilization was performed in a moderately acidified forest lake in central Sweden, to restore the autochthonous productivity. Evaluation following two years of gentle nutrient enrichment (0 08 g P · m^{-2} and 0.28 g N · m^{-2}) during summer stratification showed increased production through the entire planktonic food chain without distinct structural changes.

The results are further evaluated together with planktonic structures reported from acidified, limed, fertilized, and unperturbed oligotrophic lakes The quantitative concept used for evaluation of planktonic structure showed general differences between acidified and undisturbed oligotrophic lakes. Lakes treated with lime indicated no transition of the altered planktonic structure of acidified lakes to the efficient planktonic food chain observed in non-acidified lakes. The possibility to restore plankton communities in acidified lakes with lime alone or in combination with nutrients is discussed

Introduction

Structural changes in freshwater ecosystems resulting from planned or accidental perturbations of lakes have frequently been reported. Several experiments in lakes have also successfully shown that the planktonic food chain, nutrients — algae — herbivore — carnivore, can be roughly governed from the endpoints. Fertilization of oligotrophic lakes has thus been shown to improve food resources for planktivorous fish and to increase fish production (NELSON 1959, LE BRASSEUR et al. 1978, MILLS 1985). Results in the opposite direction have been presented from studies of fish removal in eutrophic lakes resulting in decreased algal biomass (REINERTSEN & OLSEN 1984) and also reduced nutrient concentrations (ANDERSSON et al. 1978, SHAPIRO & WRIGHT 1984). These results and others may be regarded as incentives for future successful management of freshwater productivity. Difficulties with biomanipulations have also been highlighted by these studies, e.g. instability and collapses of nutrient-stimulated food chains (REINERTSEN & LANGELAND 1982) and reduced efficiency of planktonic food chains due to fertilizer-induced algal blooms of species with low nutritional value to herbivores (PERSSON 1985). These initial experiences of biomanipulation, mostly within the field of eutrophication research, may also contribute to the restoration of acidified lakes.

The structural changes in oligotrophic lake ecosystems caused by acidification have now been well documented (ALMER et al. 1978, HAINES 1981, among others). Fish populations have initially been reduced in number because of impacts on reproduction before being finally eliminated. Diversity of phytoplankton and zooplankton communities has been reduced. Furthermore, recent results show that the concentrations of total phosphorus have successively decreased during acidification as a result of decreasing input of phosphorus from the drainage area (BROBERG & PERSSON 1984, JANSSON et al. 1986).

Liming, the generally accepted way of curing acidified lakes, has been demonstrated to restore pH, alkalinity and the reproductive conditions of fish (NYBERG et al. 1986a). Plankton has been reported to respond to liming with lowered biomass, but shows generally great similarities in temporal development before and after liming. Both increased and decreased phosphorus concentrations have been reported after liming (BROBERG 1987), but phosphorus measured as seasonal mean concentration has been found to remain at about the same level as before the lime treatment (DILLON et al. 1979, JANSSON et al. 1986). There is, thus, growing evidence that liming of the lake alone is insufficient to restore the lake to pre-acid conditions. The well-known importance of phosphorus as a factor regulating primary production, and thereby the entire ecosystem, of natural lakes has also been demonstrated to be valid in acidified as well as in limed lakes (YAN & LAFRANCE 1984, WILCOX & DECOSTA 1984). Liming, followed by gentle fertilization, might then be a reasonable combination of methods to increase the degree of restoration of acidified lake ecosystems.

This paper presents results from a full-scale experiment in which gentle fertilization of a limed lake has been carried out in order to increase the concentrations of nutrients, and thereby the planktonic productivity, but without damaging the oligotrophic nature of the ecosystem. The changes observed in the planktonic structure of the experimental lake have been evaluated together with data from other oligotrophic lakes suffering from varying degrees of planned or accidental perturbations and from undisturbed oligotrophic lakes.

Methods

The fertilization experiment was carried out in Lake Hecklan, situated in the forested region of central Sweden at 59° 53′ N, 14° 32′ E or at 664169—142838 in the system proposed by the Swedish Meteorological and Hydrological Institute (SMHI). The drainage area (1.5 km^2) has an altitude of 324—224 m above sea level. Apart from the lake, which makes up about ca. 11 %, the area is dominated by coniferous forests and bogs. The lake has one visible inlet and one outlet.

The surface area of the lake is 0.16 km^2 and the lake volume is 1.1 million cubic meters. The turnover time of the water is 1.5 to 1.8 years. The lake is dimictic and is covered by ice for about 160 days, from November 20 to May 1 Due to a wind-sheltered location at the bottom of a valley and with high surrounding vegetation, the spring and autumn circulation periods are very short, ca. one and 2—4 weeks respectively, and the typical depth of the thermocline is only 3 m

The lake water has a fairly brown colour (ca. 50 mg Pt · l^{-1}), low conductivity (ca 3.4 mS · m^{-1}) and low concentration of nutrients (Table 1). Both emergent and floating-leaved vegetation are very sparse

During the past decade, local fishermen reported lowered catches of the four species of fish present, perch, roach, brown trout and burbot, as well as of crayfish. At the end of that period, the crayfish had disappeared and the roach population showed signs typical of acid stress. Therefore, a lime treatment was proposed. Liming was performed in March 1983, when a total of 44.4 tonnes, or 2 8 tonnes · ha^{-1}, was spread on the ice. As a result of the liming, alkalinity increased from 0.06 to 0.10 meq · l^{-1} and pH from 6.2 to 6.5. The liming of the lake was thus conducted at an early stage of acidification, before the collapse of the bicarbonate buffer system.

During 1983, the sampling as well as the analyses of water chemistry, zooplankton numbers (epilimnion samples) and fish community structure were conducted by Dr. P. Nyberg at the Institute of Freshwater Research, Drottningholm, Sweden. These unpublished data, as well as samples of phytoplankton (epilimnion) and zooplankton (net hauls covering surface to bottom), were kindly placed at our disposal and constitute the background data to our fertilization experiment which was carried out in 1984 and 1985.

The fertilization programme was based on a slow and continuous release of phosphorus and nitrogen from a commercial fertilizer (Osmocote) enclosed in nine net cages. The net cages were evenly distributed over the lake surface and reached to a depth of about 1 m depth The actual release of nutrients during stratification (May—Oct.) was calculated from analysis of the residual content of the pellets. Based on laboratory experiments, the exposed amounts of pellets theorectically were calculated to add around 0.08 g P · m^{-2} and 0.28 g N · m^{-2} to the lake in both years.

Sampling was performed biweekly during May—October, using a tube sampler of 2m length. Volume-weighted composite samples representing epi- + metalimnion (0—5 m depth) and hypolimnion (5—17 m depth), respectively, were taken at four stations. Subsamples from each of the two composite samples were then taken for analysis of water chemistry, phytoplankton and zooplankton.

In order to estimate total nutrient load on the lake, water samples were also taken from the inflow creek. Area-specific runoff data were obtained from field studies in the nearby region (Grip 1982, Bergquist et al. 1984) and data on atmospheric deposition of phosphorus and nitrogen were adopted from Persson & Broberg (1985).

Sedimentation from the epilimnion was measured by using funnel traps, containing 4% formalin, placed at three stations in the hypolimnion of the lake. Before analysis of nutrient content the sedimented material was filtered through a 200 μm mesh-size net in order to remove larger migrating animals.

Total phosphorus (TP) was analysed according to Menzel & Corwin (1965), soluble reactive phosphates (SRP) were analysed on filtered water (Murphy & Riley 1962) and particulate phosphorus (PP) was collected on 0 2 μm membrane filters (Schleicher & Schüll) followed by acid digestion and photometric detection of phosphomolybdate (Ahlgren & Ahlgren 1976). Ammonium was analysed according to Chaney & Marbach (1962) and nitrate according to Wood et al. (1967). Kjeldahl nitrogen was analysed according to Jonsson (1966) and total nitrogen (TN) was calculated as the sum of Kjeldahl-N and nitrate + nitrite. Particulate nitrogen (PN) and particulate carbon (PC) were collected on pre-heated (550 °C) glassfibre filters (Whatman GF/F) and analysed by dry ignition and IR detection.

Phytoplankton cell numbers were determined in the conventional way with an inverted microscope. Biovolumes were obtained from measurements of linear dimensions of the cell and transferred to biomass assuming a cell density of 1 g · cm^{-3}. For analysis of zooplankton, 10 l of water were filtered through 40 μm net and preserved with Lugol's solution. These samples were counted in their entirety under an inverted microscope as regards numbers and eggs (loose eggs were referred to species). Crustaceans were measured by length and their biomass was calculated using length — dry weight — relationships established for nearby lakes (Olofsson in prep.) Data from these lakes were also used to obtain biomass of rotifers.

The comparative evaluation of the planktonic structure in oligotrophic lakes of different states of human disturbance was based on literature data Lakes were selected for the comparison when seasonal means ($n > 5$) of at least three of the parameters pH, TP, community biomass of phytoplankton (Bp), taxonomic composition of Bp, and community biomass of zooplankton (Bz), were available and determined with methods similar to those used in the present study. Phytoplankton composition was used to calculate the quotient, Bp (biomass of Chrysophyceae + Cryptophyceae)/Bp, which is used as an indication of the percentage of Bp with reduced or low nutritional value for filter-feeding herbivores (cf. Nauwerck 1963, Porter 1973, Persson & Johansson 1984) Another quotient, Bz/Bp, calculated on a carbon basis as Bz (in μg dw · l^{-1}) · 0.5/Bp (in μg ww · l^{-1}) 0.1, was used to indicate utilization of autochthonous algal production by planktonic herbivores

Results and Discussion

Effects of fertilization on nutrients and plankton

The effects of fertilization in Lake Hecklan on mean summer concentrations of nutrients and plankton are

Table 1. Mean summer (1 June—15 Sept.) levels of nutrients and plankton biomasses in Lake Hecklan 1983 (limed lake), 1984 (1st year of N,P-additions) and 1985 (2nd year of N,P-additions).
Chemical data for 1983 are obtained during May—October.
For the explanation of abbreviations, see chapter "Methods".

	1983			1984						1985					
	<5 m			<5 m			>5 m			<5 m			>5 m		
	n	$\bar{x}$	C.V.	n	$\bar{x}$	C.V.	n	$\bar{x}$	C.V.	n	$\bar{x}$	C.V	n	$\bar{x}$	C V.
PC (μg/l)	—	—	—	8	422	19	8	340	34	7	712	25	7	493	22
TP (μg/l)	4	6.2	30	9	10.2	23	9	7.8	17	7	10 5	30	7	9.8	20
PP (μg/l)	—	—	—	9	7.9	34	7	5 1	18	8	6.0	48	4	6 0	30
SRP (μg/l)	—	—	—	9	1 8	33	9	1 7	36	6	2.2	44	4	2.9	7
TN (μg/l)	1	370	—	9	282	9	9	334	14	7	314	9	7	320	34
PN (μg/l)	—	—	—	8	48	37	7	32	22	7	77	25	7	44	25
NO_3-N (μg/l)	1	70	—	9	10	15	9	100	20	7	19	90	5	31	143
NH_4-N (μg/l)	—	—	—	9	8	64	8	12	59	7	6	50	7	14	41
TN/TP	1	41	—	9	29	32	8	44	22	7	32	26	7	32	14
PN/PP	—	—	—	8	6.5	24	7	6.5	33	6	17	78	4	8.5	31
B_p (μgWW/l)	6	101	41	9	511	47	10	102	50	6	997	56	6	113	36
B_z (μgDW/l)	*6	50.2	60	8	95.8	78	8	55.8	37	7	160	58	7	49.3	41

* Exclusive rotifers

presented in Table 1. During the first year of fertilization TP-levels in the epilimnion were doubled in comparison with the pre-fertilization values, to about 10 µg $P \cdot l^{-1}$. The same level of TP was maintained in the epilimnion also during the second year of nutrient additions. In both 1984 and 1985, the average TP concentrations in the hypolimnion were slightly lower than those in the epilimnion. Mean concentrations of TN were not elevated and very similar in the entire waterbody during fertilization. The TN/TP ratio was well above 17, which indicates that phosphorus was the limiting nutrient for algal growth (HEALEY 1975). The epilimnetic mean levels of inorganic nitrogen and SRP were low during fertilization and most of the nutrients released were thus rapidly bound to organic molecules and particles.

The most pronounced effect of fertilization was found on Bp, which was 10 times higher in the epilimnion in the second year of nutrient additions than during the reference year. The corresponding increase of Bz was around three times.

Fate of added nutrients

Nitrogen released from the fertilization cages to the epilimnion made up about 25% of the total external load in both years (Table 2). The corresponding values for phosphorus were about 85%, which means that the phosphorus load had increased about five-fold. Measured TP in the epilimnion (Fig. 1a) increased from 5 to 15 µg $\cdot l^{-1}$ during the first weeks after the installation of the nutrient cages in 1984 and then declined continuously to 7 $\mu g \cdot l^{-1}$ in October when the nutrient cages were removed from the lake. The same seasonal development of TP was also measured during the second year of fertilization. The initial increase of about 10 $\mu g\ P \cdot l^{-1}$ corresponds to approximateley 50% of the total amount of phosphorus released during the season. The rate of nutrient release was thus much higher at the beginning of the growing season than later during summer.

Table 2. Nutrient loading and sedimentation in Lake Hecklan during stratification. Expressed as kg/lake and in % of total load (within brackets).

	Fertilization	Total input	Sedimentation
1984			
P	14.3 (83)	3.0 (17)	10.8 (62)
N	42.0 (25)	129.0 (75)	71.0 (42)
C	— —	— —	730 —
N/P	2.9	43	6.6
1985			
P	12.9 (85)	2.2 (15)	12.8 (85)
N	49.0 (26)	136.0 (74)	119.0 (64)
C	— —	— —	1190 —
N/P	3.8	6.2	9 3

The sedimentary losses of nitrogen from the epilimnion during fertilization were measured to 40—60% of the calculated input. Observed sedimentation of phosphorus was 60—80% of the total phosphorus input, or close to the amount of phosphorus added by fertilization. The main part of the phosphorus added seems therefore to have been utilized by plankton and only minor parts could have been assimilated by littoral vegetation or adsorbed in epilimnetic bottoms unless sedimentation is highly overestimated. There are, however, three reasons to believe that the observed values of sedimentation are realiable: 1. The elemental composition in epilimnic seston (Table 1) was in good agreement with the composition of sedimented material (Table 2). 2. Sedimented carbon (Table 2) in 1984 and 1985 corresponds to a mean daily primary production of 8 and 14 $mg\ C \cdot m^{-3}$, respectively. In comparison to reported mean seasonal values of 18—36 $mg\ C \cdot m^{-3}$ obtained with the 14C-method during stratification in similar forest lakes (HEYMAN 1983), the observed values are not to high. 3. Observed Bp (Table 1) and sedimented carbon (Table 2) indicate a daily carbon turnover around 0.15 during both years of fertilization, which instead is an underestimate in this type of forest lake (op. cit.).

Thus, it seems justified to conclude that added nutrients primarily were utilized by plankton and that most of the added nutrients sedimented to hypolimnetic bottoms.

Interpretation of phytoplankton development

During the pre-fertilization year, 1983, Bp was very low (Table 3) and comparable to Bp values recorded in extremely oligotrophic and stratified subarctic lakes (e.g. HOLMGREN 1983). Apart from a minimum in late June (Fig. 1b) little variation occurred during the growing season. The Bp values during 1984 and 1985 were 5 and 10 times higher, respectively, and maxima developed in the middle of the stratified period both years. The increase of Bp during fertilization is greater than what could be expected from normal year-to-year variations (e.g. RAMBERG 1979), and the early summer minimum,

Table 3 Phytoplankton community structure, taxonomic groups and major contributing species, in Lake Hecklan 1983—1985. The composition based on average seasonal community biomass in epilimnion.

Group/Taxon	Biomass (%)		
	1983	1984	1985
Cyanophyceae	3	2	1
Chrysophyceae	46	35	64
Ochromonadaceae spp	36	25	16
Uroglena sp.	4	0	9
Mallomonas akrokomos RUTTN.	3	+	+
Mallomonas cf. *crassisquama* (ASM.) FOTT	+	+	30
Bacillariophyceae	13	9	10
Melosira distans v. *alpigena* GRUN.	+	+	3
Rhizosolenia longiseta ZACH.	0	4	+
Synedra acus KÜTZ.	0	5	6
Cyclotella sp.	12	+	0
Volvocales	1	3	2
Chlorococcales	6	8	5
Oocystis sp.	3	+	+
Koliella sp.	0	5	3
Cryptophyceae	27	29	16
Cryptomonas marssonii (MARSS.) SKUJA	10	4	+
Rhodomonas lacustris PASCH. & RUTTN.	11	17	9
Katablepharis ovalis SKUJA	3	6	3
Dinophyceae	4	14	2
Gymnodinium cf. *lacustre* SCHILL.	3	10	+
Gymnodinium uberrimum (ALLM.) KOF. & Sw	+	3	+
Total Biomass $\mu g\ ww \cdot l^{-1}$	101	511	997

+ denotes presence in amounts less than 1%
0 denotes that the species was not recorded in the samples.

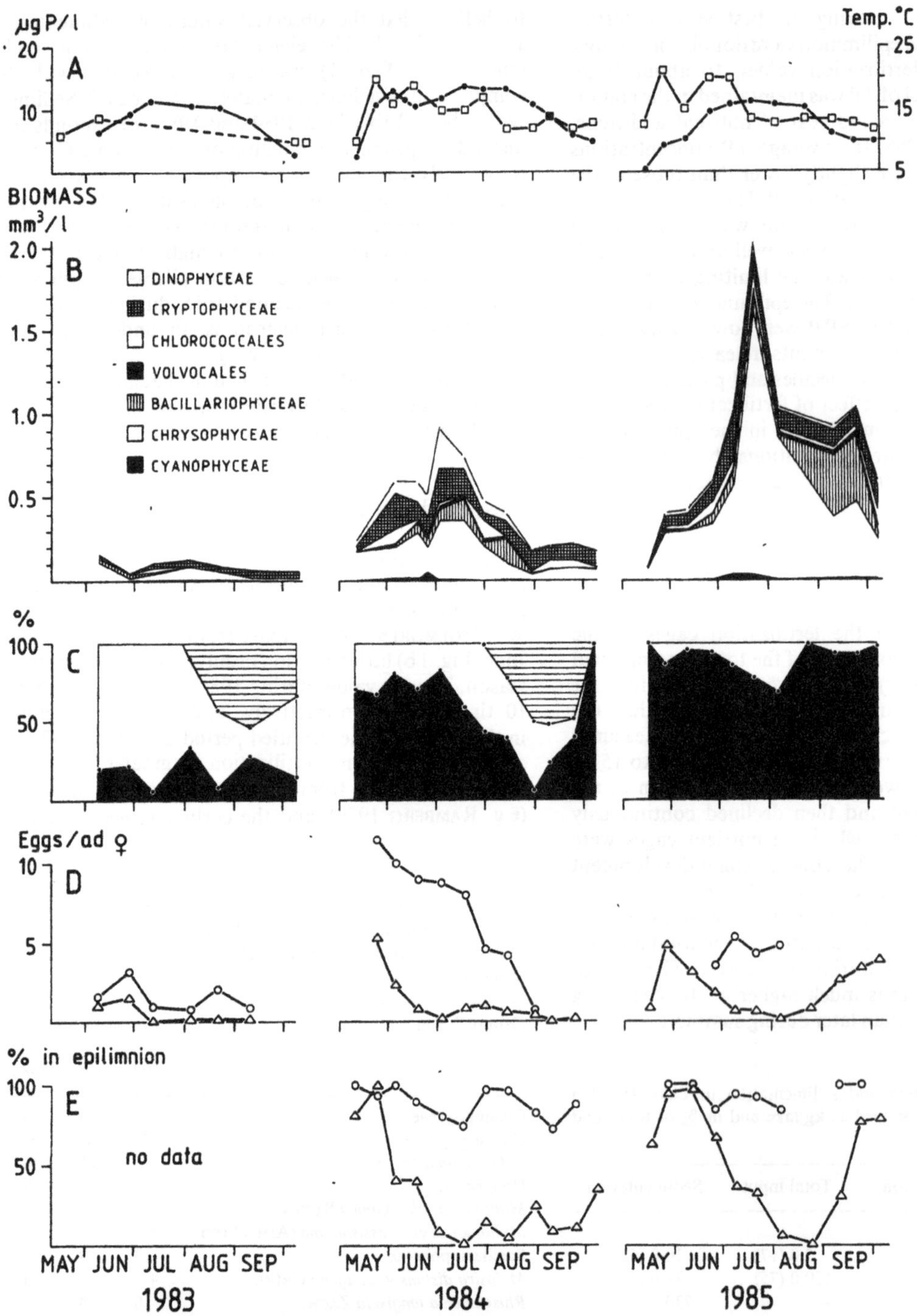

Fig. 1. Selected abiotic and biotic conditions in Lake Hecklan 1983—1985.
A: Squares denote total phosphorus (TP) and dots temperature.
B: Total phytoplankton biomass (Bp) and its composition.
C: Relative composition of *Daphnia* spp., black areas denote *D. longiremis*, white areas denote parthenogenetic females of *D. longispina*, hatched areas denote sexually mature individuals (males and females) of *D. longispina*.
D: Fecundity of *Daphnia* spp. *D. longispina* is presented with circles and *D. longiremis* with triangles
E: Vertical distribution of *Daphnia* spp, circles denote the %-age of *D. longispina* in the epilimnion, triangles denote the %-age of *D. longiremis* in the epilimnion.

which is so often pronounced (e.g. NAUWERCK 1963, RAMBERG 1979, BLOMQVIST et al. 1981), no longer exists.

Chrysophyceae, together with Cryptophyceae, dominated the composition of Bp (Table 3) during all three years. The most important chrysophyceans were small monads of the genera *Ochromonas* and *Chromulina*, except during 1985 when *Mallomonas* cf. *crassisquama* dominated. Among the cryptophyceans, there were no changes in the species composition. Among the taxonomic groups of less quantitative importance there were only minor changes during the study.

Thus, as regards the composition of phytoplankton, the fertilization of Lake Hecklan resulted in very minute changes and the phytoplankton community remained typical of non-acidic, low-alkaline and oligotrophic forest lakes (e.g. RAMBERG 1979, BLOMQVIST et al. 1981,

ROSÉN 1981). Bp, on the other hand, increased from extremely low levels commonly recorded in the kind of lakes mentioned above.

Interpretation of zooplankton community

Mean seasonal biomass of zooplankton, calculated as a weighted average for the whole lake, increased from 45 to 95 µg · l^{-1} during the fertilization study (Table 4). Compared to Bz values between 60 and 120 µg dw · l^{-1} estimated for other forest lakes in the same area (BLOMQVIST et al. 1981, OLOFSSON in prep.), Bz in lake Hecklan was low before fertilization and rose to normal levels during fertilization.

Table 4. Zooplankton community structure in Lake Hecklan 1983—1985. The composition is based on seasonal averages of volume-weighted biomass estimates representative to the whole lake.

	Biomass %		
	1983	1984	1985
Rotatoria			
Keratella hiemalis CARLIN	+	+	+
Keratella cochlearis (GOSSE)	+	1	+
Kellicottia longispina (KELLICOTT)	1	1	1
Ascomorpha sp.	+	+	+
Asplanchna priodonta GOSSE	+	+	+
Polyarthra major (BURCKHARDT)	+	+	+
Polyarthra vulgaris CARLIN	2	2	1
Conochilus unicornis ROUSSELET	+	4	3
Collotheca sp.	+	+	+
Cladocera			
Diaphanosoma brachyurum (LIEVIN)	+	+	+
Holopedium gibberum ZADDACH	1	4	2
Daphnia longispina MÜLL.	46	30	13
Daphnia cristata SARS	+	+	+
Daphnia longiremis SARS	4	5	28
Bosmina longispina (LEYDIG)	3	1	1
Bythotrephes longimanus LEYDIG	+	+	+
Copepoda			
Heterocope appendiculata SARS	5	6	5
Eudiaptomus gracilis (SARS)	21	34	35
Cyclops scutifer SARS	12	7	9
Thermocyclops oithonoides (SARS)	5	4	1
Total Biomass µg dw l^{-1}	45*	68.1	94.6

* Estimated from measured crustacean biomass in epilimnion (Table 1) and its relation to weighted community biomass.

The composition of the zooplankton community was rather stable during the study (Table 4). Rotifers constituted less than 10% of Bz in all years and the remaining Bz was made up to almost equal parts of cladocerans and copepods. However, a species shift occurred among the cladocerans from dominance of the larger *Daphnia longispina* to dominance of the smaller *Daphnia longiremis* (Fig. 1, panel c). Fecundity of both species (Fig. 1 panel d) increased during fertilization mainly as a result of the improved food conditions (Fig. 1, panel b), since adult length remained largely unchanged (Table 5), but the higher fecundity of *D. longispina* than of *D. longiremis* offers no explanation of the shift in species. During summer the fecundity of *D. longiremis* decreased simultaneously with an accumulation of the population in the hypolimnion (Fig. 1, panel d) where the concentration of algal food was much lower (Table 1). The vertical distribution of *D. longispina* during the same period was restricted to the epilimnion.

All other herbivore cladocerans (Table 4) occurred in the same way as *D. longispina* in the epilimnion. The highest abundances of these species were usually observed in May and early June when immature individuals originating from hatching resting eggs still dominated the population structures. During summer the abundances of the same species decreased to densities near or below the detection limit of the study (<0.05 ind. · l^{-1}). The biomasses of these species were also rather unimportant for the community biomass during fertilization in spite of the improved food conditions. Among the cladocerans, only *D. longiremis* developed a distinct increase of biomass during fertilization.

There are two important reasons to the differences observed; 1. The population growth of the perennial *D. longiremis* starts directly after ice-break from adult parthenogenetic females, while species initially recruiting their summer populations from hatching resting eggs start the parthenogenetic reproduction and growth considerably later. This difference was very significant in spring 1985, when unusually low water temperatures (Fig. 1, panel a) delayed juvenile development and parthenogenetic reproduction of *D. longispina* to June, at the same time as *D. longiremis* reached population maximum. 2. After the descent to the hypolimnion in June the population of *D. longiremis* showed rather constant numbers during the summer, indicating that the lowered fecundity was accompanied by a reduced elimination. In contrast to *D. longiremis*, the epilimnetic cladocerans must have experienced high mortality during the summer, as the high fecundity values did not induce expanding population sizes.

Table 5. Average length (L) and fecundity (E) of adult cladocerans in Lake Hecklan 1983—1985 (*n* denotes the numbers of adults measured and used for estimations of fecundity).

Species	Year	June—July			Aug.—Sept.		
		n	L-SD	E	*n*	L-SD	E
Daphnia	1983	30	1.79—0.13	1.67	32	1.98—0 20	0.97
longispina	1984	34	1.82—0.27	5.79	63	1.79—0 23	1.70
	1985	38	1.83—0.23	4.60	—	—	1.70
Daphnia	1983	—	—	—	33	1.11—0 05	0 10
longiremis	1984	47	1.14—0.09	0.82	42	1.16—0.08	0.30
	1985	200	1.20—0.10	1 77	33	1.17—0.09	0.89
Bosmina	1983	21	0.65—0.07	0.67	—	—	—
longispina	1984	16	0.62—0.11	2 56	—	—	—
	1985	26	0.58—0.07	2 38	—	—	—

Beside *D. longiremis*, also the copepod *Eudiaptomus gracilis* responded to the fertilizations with increased seasonal biomass. Together, these two crustaceans developed a biomass of about 11 µg dw · l^{-1} before fertilization, and during the second year of nutrient enrichment the corresponding value was about 60 µg dw · l^{-1}. The biomass of all other crustaceans together was rather unchanged, between 30 and 36 µg dw · l^{-1} in all three years.

Despite the generally very different life cycle of the cladoceran *D. longiremis* and the copepod *E. gracilis*, similarities were observed that explain their common development. The life cycle of *E. gracilis* in Lake Hecklan is dominated by one cohort with rapid growth from nauplii to copepodids IV and V during May in the epilimnion. During June, the copepodids descend to the hypolimnion where individual growth is retarded and elimination from the cohort is unimportant. The copepodids attain adulthood during the autumn circulation, but reproduction does not occur until April—May in the following year. The common features of these two crustaceans that separate them from the other crustaceans, as regards development of biomass during fertilization, are thus a low mortality during summer and a high capacity for rapid growth just after ice-break.

Increasing biomass during nutrient enrichment was also observed in one rotifer species, the colony-forming *Conochilus unicornis*, which became the dominant rotifer species (Table 4). However, all the solitary rotifers showed higher levels of fecundity during fertilization than during the reference year (Table 6). Especially during early summer 1985 the rotifer fecundity was high compared to observed population growth, which indicated a pattern of mortality similar to that of epilimnic cladocerans. The truly epilimnic *C. unicornis* developed population maxima during summer without any signs of important mortality, and consequently not even the features found to benefit development of crustacean biomass offer any explanation of the deviating population dynamics of this rotifer species. Almost exactly the same response of rotifers to whole lake fertilization was reported earlier by PERSSON (1978) who, after a thorough population analysis, found that increased longevity during fertilization was instrumental in creating the dominance of *C. unicornis* among the rotifers.

Table 6. Rotifer fecundity in Lake Hecklan 1983—1985.

Species	Year	June—July			Aug—Sept.		
		n	*x*	SD	*n*	*x*	SD
Keratella cochlearis	1983	3	0.25	0.08	—	—	—
	1984	4	0.36	0.13	4	0.08	0,03
	1985	4	0.19	0.08	4	0.25	0,05
Kellicottia longispina	1983	3	0.13	0.03	3	0.19	0.03
	1984	4	0.45	0.09	4	0.19	0.14
	1985	4	0.40	0.26	4	0.43	0.04
Polyarthra vulgaris	1983	3	0.30	0.12	3	0.20	0.07
	1984	4	0.74	0.22	4	0.33	0.11
	1985	4	0.35	0.06	4	0.27	0.09

Thus, during fertilization, the zooplankton community in Lake Hecklan was affected by increased elimination rates of species inhabiting the epilimnion with exception of *C. unicornis*. Unusally rich year classes of the dominating fish species, perch, developed both 1983 and 1985 (NYBERG unpubl.). Furthermore, stomach analysis showed that also the older age-classes of perch changed their diet from one dominated by benthic food items towards one almost exclusively composed of planktonic crustaceans during fertilization (op. cit.). Increased fish stock, as well as the changed feeding habits of perch, was probably responsible for the observed pattern of zooplankton elimination.

Visually dependent planktivores are well known to have a large structuring effect on zooplankton (HRBÁČEK

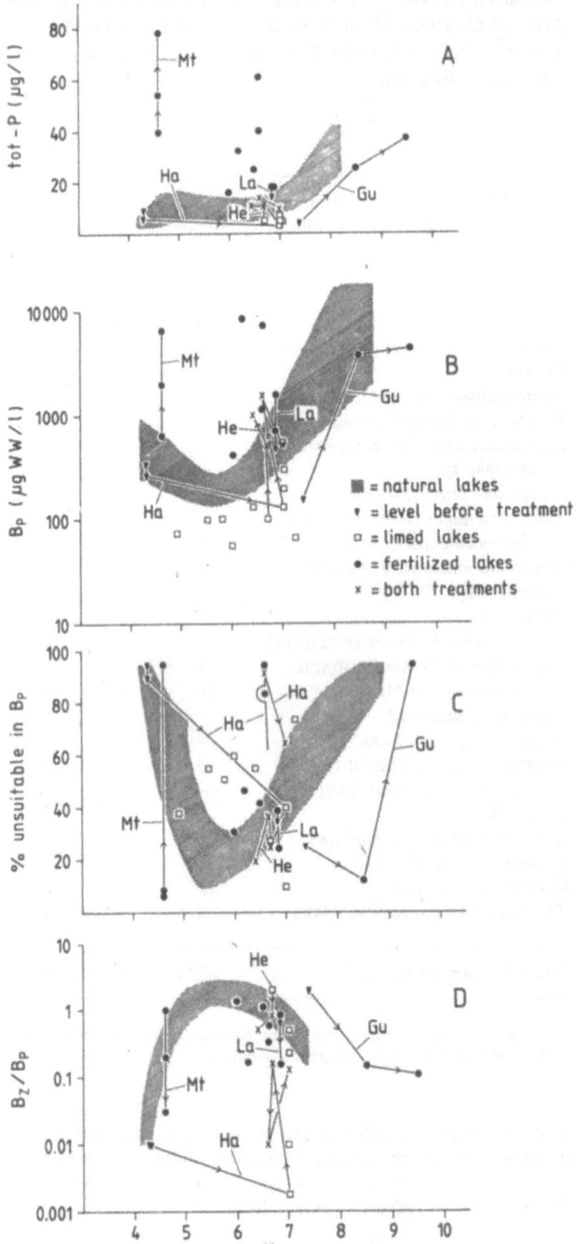

Fig. 2. Planktonic structure of oligotrophic lakes. Generalized plots of TP (A), Bp (B), %-age of unsuitable algae in Bp (C) and Z/P-ratios (D), versus pH. Further information in text.

The selected data sets have been gathered from BAALSRUD et al. (1985), BLOMQVIST et al. (1981), BLOMQVIST (unpubl.), BOSTROM & PETTERSSON (1977), DILLON et al (1979), HOLMGREN (1983), LANGELAND & REINERTSEN (1982), LARSSON, P (1978), LARSSON, S (unpubl.), NAUWERCK (1963), OLOFSSON, H (in prep.), PERSSON & JOHANSSON (1984), RAMBERG (1979) and YAN & LAFRANCE (1984).

Abbreviations Gu — Lake Gunillajaure, Ha — Lake Hannah, He — Lake Hecklan, La — Lake Langvatn, Mt — Lake Mountaintop

et al. 1961). The species composition of zooplankton in Lake Hecklan, however, still remains intact after two years of fertilization. Only the largest herbivore in plankton, *D. longispina*, showed typical signs of predatory overexploitation. The increased predation pressure on the other epilimnic herbivores vulnerable to fish predation was balanced by increased reproduction, while herbivores less susceptible to fish predation were responsible for the increased zooplankton biomass during fertilization.

The planktonic structure in acidified, limed, fertilized and undisturbed oligotrophic lakes

Figure 2 was compiled in an attempt to generalize the planktonic structure of oligotrophic lakes. Quantitative data on important plankton characteristics are plotted versus pH in order to distinguish structural changes caused by acidification, liming or fertilization.

In non-manipulated lakes, TP (Fig. 2a) decreases with decreasing pH, and in the pH-interval 5—7 TP ranges between 5 and 15 $\mu g\ P \cdot l^{-1}$. In acidified lakes with pH below 5, as well as in limed lakes, TP is usually lower and similar to the TP concentrations reported from the ultra-oligotrophic and subarctic Lake Gunillajaure (Gu in Fig. 2, Persson 1982). Two of the limed lakes, Lake Hecklan (He in Fig. 2, this study) and Hannah Lake (Ha in Fig. 2, Yan & Lafrance 1984), show levels of TP after fertilization comparable to those of non-manipulated lakes. The fertilization of Lake Langvatn (La in Fig. 2, Reinertsen & Langeland 1982) resulted in slightly higher levels, while the nutrient enrichment of the acidified Mountaintop Lake (Mt in Fig. 2, Yan & Lafrance 1984) as well as all other experimentally fertilized lakes included in Fig. 2 have resulted in TP levels around $20\ \mu g\ P \cdot l^{-1}$ or in most cases considerably higher.

Like TP, also Bp decreases with decreasing pH (Fig. 2b), but in lakes with pH below 5 Bp again is higher in spite of the low TP concentrations. Bp is usually lower in limed lakes than in acidified lakes. A subsequent fertilization of limed lakes will increase Bp in the same way as in other lakes regardless of pH at the time of nutrient enrichment. It should be noted though, that neither the load nor the resulting concentration of nutrients are in direct proportion to response of Bp. Nonetheless fertilization resulting in high TP levels ($> 20\ \mu g\ P \cdot l^{-1}$) usually causes changed algal composition and reduced diversity, while more moderate nutrient additions like those made in Lake Langvatn and Lake Hecklan lead to unchanged algal composition.

Natural oligotrophic lakes with a pH between 5 and 7 generally show algal species composition with few components considered as unfavourable food for herbivore zooplankton (Fig. 2c) (cf. chapter "Methods"). These unfavourable algae constitute a much more important part of phytoplankton biomass in both acidified lakes and alkaline lakes with high nutrient contents. Several of the limed lakes have also a composition dominated by these algal species. However, the results from Hannah Lake show that liming might lead to reduction of unfavourable algae and a changed composition towards that in non-acidified oligotrophic lakes.

In lakes with a minimum of unfavourable algae there is also a high efficiency in the planktonic food chain, which is indicated by a Bz/Bp-ratio around 1.0 (Fig. 2d). Acidified lakes below pH 5 show Bz/Bp-ratios 10—100 times lower, while limed lakes show a wide range of Bz/Bp. The very different response to liming as well as subsequent fertilization is clearly demonstrated by the different development of Bz/Bp in Hannah Lake and Lake Hecklan. The fertilization of the severely acid- and metal-disturbed Hannah Lake resulted in improved Bz/Bp-ratios, but even after three years of fertilization the lake was still not restored to non-acidified conditions. All other fertilization experiments show a reduction of Bz/Bp. In Lake Hecklan and also in Lake Langvatn, which were enriched with low loads of nutrients, the lowered Bz/Bp-ratios were observed despite almost unaltered phytoplankton communities. Increased predation by planktivorous fish was the factor responsible for changed relations between herbivore and algal biomass in both these lakes. Lowered Bz/Bp-ratios due to changed algal composition are observed when lakes are exposed to excessively high loads of nutrients, an example being the fertilization of the fishless Lake Gunillajaure.

Thus, there are general differences between acidified lakes and natural oligotrophic lakes as regards planktonic structure. Analysed lime treatments indicate no common transition of the disturbed structure backwards to the efficient planktonic food chain observed in non-acidified oligotrophic lakes. Further quantitative data including seasonal plankton development before and after treatment are required before more general conclusions on the effects of lime as well as subsequent fertilization can be made.

General discussion

The productivity of oligotrophic lakes susceptible to acidification must primarily be governed by autochthonous processes, since the kind of lakes that easily become acidified are those in the upper parts of a river system having low water turnover. When not only the lakes (e.g. Schindler et al. 1985) but also their drainage areas become acidified, a reduction in productivity is observed together with disturbances in the planktonic structure. These effects are generally explained by increased inflow of hydronium- and metal-ions (Overrein et al. 1980, Dillon et al. 1984, Campbell & Stokes 1985). In most cases lime treatments have been found to reduce the elevated concentrations of these toxic substances enough to permit reintroduction of fish. Exceptions reported are the limed lakes in Ontario, Canada, where the concentrations of metals were still lethal to fish and sublethal to zooplankton at pH values above six (Yan & Lafrance 1984). The pH and the degree of metal pollution (especially aluminium) are undoubtedly factors of importance in the successful restoration of all trophic levels in acidified lakes. However, despite successful lime treatments in those respects, the data set in Fig. 2 indicates that the disturbances to the planktonic structure and to the productivity of acidified lakes still remain after liming.

The low concentrations of phosphorus in acidified lakes have been noted as another factor responsible for the observed disturbances to biota (Almer et al. 1978, Dillon et al. 1979). The hypothesis that due to soil processes in the watersheds the input of phosphorus to lakes in acidified areas generally becomes reduced (Jansson et al. 1986), indicates that more attention should be paid to the nutrient situation in acidified lakes after liming. Observations on reduced phosphorus levels in a large number of acidified and limed lakes in south and middle Sweden during the last decade, by an average of about 50% (NYBERG et al. 1986b) support the hypothesis by JANSSON et al. (op. cit.). Thus, in addition to lime treatment additions of phosphorus also seem to be necessary when the intention is to restore the acidified lakes to pre-acid conditions.

Published experiences from whole-lake treatments of acidified lakes using combinations of base and nutrients are mainly restricted to those from the Sudbury area in Canada (YAN & LAFRANCE 1984), in which very high concentrations of metals did not permit reconstruction of the planktonic structure. The present lake, Hecklan, was limed at a very early stage of acidification compared to the Sudbury lakes, and the treatments were conducted before any serious structural changes to the plankton occurred. Sofar the first two years of observed response in Lake Hecklan are promising, with indications of increased production through the entire planktonic food chain. The technique with continuous release of a low nutrient load is obviously one reason for the successful fertilization, as fertilizations resulting in high nutrient concentrations are usually followed by undesired changes in the phytoplankton communities (cf. Fig. 2).

The only available study of fertilization of an acid and unlimed lake (Lake Mountaintop in Fig. 2) unfortunately lacks data on the planktonic structure before treatment, but a plankton community typical of unacidified oligotrophic lakes developed during the first year of fertilization. The probability that fertilization alone should improve the planktonic structure is, however, contradicted by the results of bag-experiments conducted by HAVENS & DECOSTA (1986). These authors show that fertilization of an artifically acidified plankton assemblage, consisting of low herbivore biomass and algal biomass dominated by chlorophyceans and dinophyceans, resulted in increased biomass of dinophyceans and a further decrease in the Bz/Bp-ratio. Thus, the prescence of an undamaged plankton community in Lake Hecklan at the moment of fertilization might have been an even more important factor to the successful treatment than the applied fertilization technique.

As regards the possibility to restore the inefficient planktonic structure of acidified lakes by liming, the results are rather conflicting. The Norwegian liming project concluded that liming of acidified lakes caused no clear-cut changes in the phytoplankton communities (BAALSRUD 1985). In summarizing the results on liming of lakes in Ontario, Canada, YAN & DILLON (1984) report that although chrysophytes replaced dinophyceans, the species diversity remained low. HÖRNSTRÖM & EKSTRÖM (1986), on the other hand, conclude from studies on the Swedish west coast that composition of both phyto- and zooplankton changed markedly after liming and became similar to that in unacidified lakes. The reported divergences are probably caused by different liming methods together with varying degrees of acid destruction of the planktonic structure before liming, but the absence of a general concept for evaluation of planktonic structure is also of importance.

The quantitative evaluation according to Fig. 2, although its precision suffers from too few data sets, indicates that the primary obstacle to restoration of autochthonous production in acidified lakes is the difficulty to recreate the efficient plankton structure of unacidified lakes. In this respect, neither the study in Lake Hecklan nor the other reported treatments of acidified lakes provide information on a suitable technique for successful restoration. The method used in Lake Hecklan shows that nutrient-starved but still intact plankton communities in lakes under the influence of acid precipitation can be restored as regards the level of autochthonous production and thus lead to increases in the biological resistance to acidification. Consequently, the method is rather preservative than restorative as long as the possibilities to transfer plankton structures in various degree of acid destruction towards the efficient planktonic food chain in undisturbed oligotrophic lakes remain unknown.

Zusammenfassung

Die Kalkung und die nachfolgende Düngung eines mäßig sauren Sees (Hecklan-See) im zentralen Schweden führte zu einem Anstieg der Produktion innerhalb der gesamten planktischen Nahrungskette, ohne daß es zu strukturellen Änderungen kam. Es wird diskutiert, ob zur Restaurierung planktischer Gemeinschaften saurer Seen die Zugabe von Kalk allein oder in Verbindung mit Nährstoffen erfolgen sollte.

Acknowledgements

We wish to express our gratitude to Doz. Dr. JÜRGEN BENNDORF and Prof. BIRGER PEJLER for valuable comments on the manuscript.

References

AHLGREN, I., & AHLGREN, G. (1976): Vattenkemiska analysmetoder. Inst. of Limnol., Uppsala Univ, 80 p. (in Swedish).

ALMER, B., DICKSON, W., EKSTRÖM, C., & HÖRNSTRÖM, E. (1978): Sulfur pollution and the aquatic ecosystem. In: J. O. NRIAGO (ed.), Sulfur in the environment: part II. pp. 271—311. New York.

ANDERSSON, G., BERGGREN, H., CRONBERG, G., & GELIN, C. (1978): Effects of planktivorous and bentivorous fish on organisms and water chemistry in eutrophic lakes. Hydrobiologia **59**, 9—15.

BAALSRUD, K. (1985): Kalking av surt vann. Kalkingprosjektet sluttrapport Direktoratet for Vilt og Ferskvannsfisk, Trondheim. 145 p. (in Norwegian).

BERGQUIST, B., LUNDIN, L., & ANDERSON, A. (1984): Hydrologiska och limnologiska konsekvenser av skogs- och myrdikning. Siksjöbäcksomradet. Report Limnol. Inst., Uppsala Univ. B4, 143 p. (in Swedish).

BLOMQVIST, P., HEYMAN, U., & GRUNDSTRÖM, R. (1981): The structure of the pelagic ecosystem in Lake Siggeforasjön. Scripta Limnologica Upsaliensia, Scripta A **522**, 79 p.

BOSTRÖM, B., & PETTERSSON, K. (1977): The spring development of phytoplankton in Lake Erken. Freshwat. Biol. **7**: 327—335.

BROBERG, O. (1987): Nutrients responses to the liming of Lake Gårdsjön. Hydrobiologia **150**: 11—24.

— & PERSSON, G. (1984): External budgets for phosphorus, nitrogen and dissolved organic carbon for the acidified Lake Gårdsjön. Arch. Hydrobiol. **99**, 160—175.

CAMPBELL, P. G. C., & STOKES, P. M. (1985): Acidification and toxicity of metals to aquatic biota. Can. J. Fish. Aquat. Sci. **42**, 2034—2049.

CHANEY, A. L., & MARBACH, E. P. (1962): Modified reagents for the determination of urea and ammonia. Clin. Chem. **8**, 130 to 132.

DILLON, P. J., YAN, N. D., SCHEIDER, W. A., & CONROY, N. (1979): Acidic lakes in Ontario, Canada: Characterization, extent and responses to base and nutrient additions. Arch. Hydrobiol., Beih. Ergebn Limnol. **13**, 317—336.

— YAN, N. D., & HARVEY, H. H. (1984). Acidic deposition: Effects on aquatic ecosystems. CRC **13** (3), 167—194.

GRIP, H. (1982): Waterchemistry and runoff in forest streams at Kloten. UNGI Report 58. Uppsala, Sweden, 144 p.

HAINES, T. A. (1981): Acid precipitation and its consequences for aquatic ecosystems. A Review. Trans. Amer Fish. Soc. **110**, 669—707.

HAVENS, K. E., & DECOSTA, J. (1986): A comparison of phytoplankton responses to nutrient additions in acidic and circumneutral pH lakewater. Hydrobiologia **137**, 211—222.

HEALY, F. P. (1975): Physiological indicators of nutrient deficiency in algae. Fish. Mar. Serv. Res. Dev. Tech. Rep **585**, 30 p.

HEYMAN, U. (1983). Relations between production and biomass of phytoplankton in four Swedish lakes of different trophic status and humic content. Hydrobiologia **101**, 89—104.

HOLMGREN, S. (1983): Phytoplankton biomass and species composition in natural, fertilized and polluted subartic lakes. Acta Universitatis Uppsaliensis **674**, 160 p.

HRBÁČEK, J., DVORAKOVA, M., KORINEK, V., & PROCHÁZKOVÁ, L. (1961). Demonstration of the effect of the fish stock on the species composition of zooplankton and the intensity of metabolism of the whole plankton association. Verh. Int. Verein. Limnol. **14**, 192—195.

HÖRNSTROM, E., & EKSTRÖM, C. (1986): Acidification and liming effects on phyto- and zooplankton in some Swedish west coast lakes Swedish Environmental Protection Board, Report **1864**, 108 p.

JANSSON, M., PERSSON, G., & BROBERG, O. (1986): Phosphorus in acidified lakes · The example of Lake Gårdsjön, Sweden Hydrobiologia **139**: 81—96.

JÖNSSON, E. (1966): The determination of Kjeldahl nitrogen in natural water Vattenhygien **1966**, 10—14.

LANGELAND, A., & REINERTSEN, H. (1982): Interactions between phytoplankton and zooplankton in a fertilized lake. Holarct. Ecol. **5**, 253—272.

LARSSON, P. (1978): The life cycle dynamics and production of zooplankton in övre Heimdalsvatn. Holarct. Ecol. **1**, 162—218.

LE BRASSEUR, R. J., MC ALLISTER, D., BARRACLOUGH, W. E., KENNEDY, O. D., & STEPHENS, K. (1978): Enhancement of sockeye salmon (*Oncorhynchus nerka*) by lake fertilization in Great Central Lake: Summary report. J. Fish. Res. Bd. Can. **35**, 1580 to 1596.

MENZEL, D. W., & CORWIN, N. (1965): The measurements of total phosphorus in sea-water based on liberation of organically bound fractions by persulfate oxidation. Limnol. Oceanogr. **10**, 280—282.

MILLS, K. H. (1985): Responses of lake whitefish (*Coregonus clupeaformis*) to fertilization of lake 226, the experimental lakes area Can J. Fish. Aquat. Sci **42**, 129—138.

MURPHY, J., & RILEY, J. P. (1962) · A modified single solution method for determination of phosphate in natural waters. Anal Chim Acta **27**, 31—36.

NAUWERCK, A. (1963): Die Beziehungen zwischen Zooplankton und Phytoplankton im See Erken. Symb. Bot. Upsal. **17**, 1—163

NELSON, P. R (1959): Effects of fertilizating Bare Lake, Alaska, on growth and production of red salmon (*Oncorhynchus nerka*). U.S. Fish. Wildl. Serv. Fish. Bull. **60**, 59—86.

NYBERG, P., APPELBERG, M., & DEGERMAN, E. (1986a): Effects of liming on crayfish and fish in Sweden. Proceed. Int. Conf. on acid precipitation. Water, Air and Soil pollut. **31**: 669—689.

— DEGERMAN, E., EKSTRÖM, C., & HÖRNSTRÖM, E. (1986b). Acid-sensitive arctic char, (*Salvelinus alpinus*), lakes in southern and central Sweden. Inf. Sötvattenlab. Drottningholm **6**, 240 p (in Swedish, with English summary).

OVERREIN, L. N., SEIP H M., & TOLLAN, A. (1980) Acid precipitation — effects on forest and fish Final report in SNSF-projects 1972—1980. FR 19/80. SNSF project, Oslo-Ås, 175 p.

PERSSON, G. (1978): Experimental lake fertilization in the Kuokkel area, northern Sweden The response by the planktonic rotifer community. Verh. Int. Verein. Limnol. **20**, 875—880

— (1982) Losses of isotopic label upon liquid preservation of zooplankton in feeding and assimilation studies. Arch. Hydrobiol **94**, 502—519.

— (1985). Clearance rates of crustacean macrofiltrators: The nature of in situ rate depressions in a fertilized oligotrophic lake in the Kuokkel area, northern Sweden. Int. Rev. ges. Hydrobiol **70**: 335—358.

— & BROBERG, O. (1985): Nutrient concentrations in the acidified Lake Gårdsjön: The role of transport and retention of phosphorus, nitrogen and DOC in watershed and lake. Ecol. Bull. (Stockholm) **37**, 158—175.

— & JOHANSSON, J. (1984): Experimental lake fertilization in the Kuokkel area, northern Sweden: Zooplankton responses, and the precision of population estimates. (Manuscript).

PORTER, K. G. (1973): Selective grazing and differential digestion of algae by zooplankton. Nature **244**, 179—180.

RAMBERG, L. (1979): Relations between phytoplankton and light climate in two Swedish forest lakes. Int Rev. ges. Hydrobiol. **64**, 749—782.

REINERTSEN, H., & LANGELAND, A. (1982): The effects of a lake fertilization on the stability and material utilization of a limnetic ecosystem. Holarct. Ecol. **5**, 311—324.

— & OLSEN, V. (1984): Effects of fish elimination on the phytoplankton community of a eutrophic lake. Verh Int. Verein. Limnol. **22**, 649—657.

ROSÉN, G. (1981): Tusen sjöar Växtplanktons Miljökrav. Liber. Stockholm, 120 p.

SHAPIRO, J., & WRIGHT, D. I. (1984): Lake restoration by biomanipulation: Round Lake, Minnesota, the first two years. Freshwat. Biol. **14** (4), 371—383.

SCHINDLER, D. W., MILLS, K. H., MALLEY, D. F., FINDLAY, D. L., SHEARER, J. A., DAVIES, I. J., TURNER, M. A., LINSEY, G. A., & CRUIKSHANK, D. R. (1985): Long-term ecosystem stress: the effect of years of experimental acidification on a small lake. Science **228**, 1395—1401.

WILCOX, G. R., & DECOSTA, J. (1984): Bag experiments on the effect of phosphorus and base addition on the algal biomass and species composition of an acid lake. Int. Rev. ges. Hydrobiol. **69**, 173—199

WOOD, E. D., ARMSTRONG, F. A. J., & RICHARDS, F. A. (1967): Determination of nitrate in sea-water by cadmium-copper reduction to nitrite. J. Mar Biol. Ass. U. K. **47**, 23—31

YAN, N. D., & DILLON, J. P. (1984): Experimental neutralization of lakes near Sudbury, Ontario. In: NRIAGU, J (ed.), Environmental impact of smelters. New York, London, pp. 417—456.

— & LAFRANCE, C (1984) · Response of acidic and neutralized lakes near Sudbury, Ontario, to nutrient enrichment In · NRIAGU, J. (ed.), Enviromental impact of smelters. New York, London, pp 458—521

Received Jannuary 5, 1987

Address of the authors ·

H. OLOFSSON
P. BLOMQVIST
H. OLSSON
O. BROBERG
Institute of Limnology
Uppsala University
Box 557
S-751 22 Uppsala
Sweden

Limnologica (Berlin) 19 (1), 37–43, Juli 1988

Technische Universität Dresden, Sektion Wasserwesen, Bereich Hydrobiologie, Dresden, DDR

Analyse des Nahrungskonsums von Barschen (*Perca fluviatilis*) in der Talsperre Bautzen

An Investigation of the Food Consumption in Perch (*Perca fluviatilis*) in the Bautzen Reservoir

Von UTE KÖPKE, HEINZ SCHULTZ, RAYMOND JARCHOW, UWE HORNIG & JENS PENIG

Mit 7 Abbildungen und 5 Tabellen

Key words Food preference, Size selection, *Daphnia*, *Leptodora*, Daily food ration, Feeding pressure

Abstract

In the frame of a long-term experiment in biomanipulation, food composition in perch of 5–20 cm length was studied in the period 1980–1982, and daily ration in perch of 15–20 cm length was determined between May and October 1981 by means of field observations (on the base of gillnet catches) and experiments on evacuation rates accomplished in the reservoir – Although main stomach content was represented by *Daphnia*, a preference of larger prey items such as fishes, larvae of insects and *Leptodora* could be pointed out Copepods were only eaten when all other food organisms were rare. A dependence of consumption for *Daphnia* on food supply is described by a hyperbolic function. Daily ration was determined as 0.2–1 3 g per fish with a weight of 40–100 g. Although these values are underestimated by about 40% or possibly up to 80%, feeding pressure of perch on *Daphnia* decreased by more than 50% in comparison with typical years during the pre-biomanipulation period (1978 and 1979). The number of young fish eaten by the perch stock in 1981 corresponds to the amount representative for an outstanding perch year-class in the reservoir

Einleitung

Die meisten Seen der DDR sind auf Grund einer verhältnismäßig geringen Tiefe und eines nur schwachen Wasserdurchsatzes sehr anfällig gegenüber erhöhten Belastungen an Pflanzennährstoffen. Es gibt zur Zeit jedoch kaum noch Einzugsgebiete, die nicht starker industrieller oder landwirtschaftlicher Nutzung unterliegen, so daß sehr viele der stehenden und gestauten Gewässer bereits Symptome einer fortgeschrittenen Eutrophierung aufweisen. Durch ein Überangebot an Pflanzennährstoffen kommt es zu einer Massenentwicklung von Phytoplanktern und damit zu einer starken Beeinträchtigung der Wasserqualität. Dieser Erscheinung gilt es entgegenzuwirken. Durch gezielte Eingriffe in die Gewässerökosystemstrukturen kann in dieser Hinsicht bereits eine Verbesserung der Wasserbeschaffenheit herbeigeführt werden.

Die hier vorgestellten Arbeiten sind in ein Langzeitexperiment zur Biomanipulation in der hypertrophen Talsperre Bautzen eingebettet (vgl. BENNDORF et al. (1988). Ziel der Untersuchungen war eine Analyse der Barschpopulation in der Phase zwischen vorwiegend planktivorer Lebensweise und bereits möglicher piscivorer Ernährung. Der Zeitraum der Untersuchungen (1981/82) fällt genau mit dem Beginn des Wirksamwerdens der Biomanipulation (1981) zusammen und ist daher von besonderem Interesse.

Material und Methoden

Detaillierte Nahrungsuntersuchungen wurden an Barschen (*Perca fluviatilis*), die 1980–81 in der Talsperre Bautzen gefangen wurden, durchgeführt. Der Fang erfolgte mit Kiemennetzen von 24 und 28 mm Maschenweite, die untersuchten Barsche hatten somit eine Länge von 15–20 cm und waren im Durchschnitt 4 Jahre alt Untersuchungen im Jahre 1982 erstreckten sich auf den Längenbereich von 10–15 cm. Trotz entsprechender Maschenweiten blieben die Fänge dieser Längengruppen sehr gering. Die Fänge wurden in den meisten Fällen eingefrostet, teilweise aber auch an Ort und Stelle bearbeitet bzw durch das Einspritzen von 10%igem Formalin in die Leibeshöhle für kurze Zeit konserviert. Während der weiteren Bearbeitung wurden Magen und Darm herauspräpariert und zur Konservierung in 6%iges Formalin überführt sowie für die Altersbestimmung Schuppen und Operkulum entnommen.

Von den verschiedenen Möglichkeiten der Nahrungsanalyse (vgl. WINDELL 1970) wurden die numerische und die gravimetrische Methode angewendet

- das Frischgewicht des Mageninhalts wurde auf 1 mg genau bestimmt,
- der Mageninhalt wurde aufgeschwemmt und seine Bestandteile wurden unter dem Stereomikroskop gezählt

Um exakte Aussagen über den Fraßdruck der Barsche auf die Zooplanktongemeinschaft machen zu können, wurde die Aufnahme dieser Nährtiere in Tageskonsumtionen quantifiziert und in Beziehung zu biotischen (Nahrungsangebot) sowie abiotischen (Temperatur) Faktoren betrachtet Der Tageskonsum wurde im wesentlichen im Zeitraum von Mai bis Oktober 1981 ermittelt Die Bestimmung erfolgte in Freilanduntersuchungen durch Ermittlung der Mageninhaltsmenge (zu verschiedenen Tageszeiten zwischen 5 und 21 Uhr) und der Verdauungsrate. Nach Möglichkeit wurde alle 3–4 Stunden gefischt, ein Teil des Fanges sofort getötet und ein Teil in einen aus Gaze von 40–70 µm Maschenweite gefertigten und 2,5 m³ fassenden Verdauungskäfig eingesetzt, der in der Talsperre verankert war

Der entsprechende Tageskonsum ergibt sich somit aus

$$\sum_1^n C_t = S_2 - S_v \qquad (1)$$

oder

$$\sum_1^n C_t = S_2 - S_1 + D \qquad (2)$$

wobei $$D = S_1 - S_v \qquad (3)$$

mit C_t – Konsumtion im Zeitintervall $t_1 - t_2$, S_1, S_2 – mittlere Mageninhalte zur Zeit t_1 und t_2, S_v – mittlere Mageninhalt der Fische aus dem Verdauungsversuch, D – Nahrungsmenge, die verdaut wurde

Bei geringen Fängen war es teilweise nicht möglich, einen Verdauungsversuch durchzuführen, so daß eine allgemeingültige Beziehung für den Verdauungsvorgang aufgestellt wurde

$$S_v = 100\, e^{-(-0,62 + 0,03T)t} \qquad (4)$$

Damit ergibt sich für die Konsumtion im Zeitintervall $t - t_0$ analog THORPE (1977)

$$C_t = S_1 - S_0\, e^{-(-0,62 + 0,03T)t} \qquad (5)$$

mit $S_{0.1}$ – Mageninhalt zum Zeitpunkt $t_{0\,1}$, S_v – Mageninhalt nach Verdauung im Zeitintervall $t_0 - t_1$

Die mit einer linearen Gleichung beschriebene Temperaturabhängigkeit der Verdauungsrate (R) gilt nur für den untersuchten Temperaturbereich von 10—20 °C. Bei Extrapolation wird eine sigmoide Kurve die Abhängigkeit besser beschreiben, da die Verdauung selbst bei niedrigen Temperaturen nie Null ist und mit steigender Temperatur einem Maximalwert zustreben wird.

Eine Berechnung der Konsumtion nach THORPE (1977) (s. Glg. 5) berücksichtigt nicht, daß alle Vorgänge der Nahrungsaufnahme und der Verdauung gleichzeitig ablaufen. Tatsächlich müßte gelten

$$\frac{ds}{dt} = F - RS \qquad (6)$$

ELLIOT & PERSSON (1978) ermittelten den Mageninhalt zur Zeit t unter Beachtung beider Prozesse:

$$S_t = S_0\, e^{-Rt} + \int_0^t F\, e^{-Rt}\, dt \qquad (7)$$

Für die Nahrungsaufnahme gilt somit:

$$C_t = F_t = \frac{(S_t - S_0\, e^{-Rt})\, Rt}{1 - e^{-Rt}} \qquad (8)$$

mit: F_t — Nahrungsaufnahme, $S_{0,t}$ — Mageninhalt zum Zeitpunkt t_0 bzw t_n.

Für die Juli-Probennahme wurde auch eine Anwendung der ELLIOT-PERSSON-Methode für möglich erachtet Der Vergleich der Ergebnisse der beiden o. g. Methoden weist eine um 38 % höhere Tageskonsumtion nach der ELLIOT-PERSSON-Methode aus.

Da nicht zu allen Probenahmeterminen das Material in kontinuierlichen Abständen vom Morgen bis zum Abend entnommen werden konnte, wurde für jene Fälle (Mai und August) versucht, die Tageskonsumtion an Daphnien und *Leptodora* über mittlere Mageninhalte mittels der Methode von BAJKOV (1935) (zit. in ELLIOT & PERSSON 1978) zu kalkulieren:

$$D = \bar{A}\left(\frac{24}{n}\right) \qquad (9)$$

mit: D — Tageskonsum; $\bar{A}$ — mittlerer Mageninhalt; n — Zeit bis zur völligen Verdauung.

Dieser mittlere Mageninhalt wurde aus den Ergebnissen der vorliegenden Fänge berechnet. Die Zeit bis zur volligen Verdauung der Nahrung wurde unter der Voraussetzung bestimmt, daß die Verdauung exponentiell von der Zeit abhängt und eine Beziehung zur Temperatur besteht. Die Zeitspanne n wird als die Zeit aufgefaßt, in der 99% der Nahrungsmenge verdaut werden (ELLIOT 1972):

$$n_{99\,\%} = \frac{\ln 100}{R} = \frac{4{,}6052}{R} \qquad (10)$$

Nach den in Verdauungsversuchen ermittelten Verdauungsraten von 0,22 für eine Temperatur von 16 °C und ungefähr 0,1 für 12 °C ist n = 20,9 bzw. 46,0 Stunden.

Die BAJKOV-Methode wurde zunächst an Daten, mit denen bereits die Konsumtion nach THORPE ermittelt wurde, getestet. Der Vergleich der mit beiden Methoden berechneten Konsumtionen zeigt, daß die nach BAJKOV berechnete Konsumtion um 45—68 % niedriger liegt Unter Einbeziehung des Magenfüllungsgrades (MG, Einstufung nach einer 6-teiligen Skala) ließ sich ein Umrechnungsfaktor (F) von BAJKOV- auf THORPE-Werte ermitteln:

$$F = 3{,}95 - 0{,}69\ \mathrm{MG} \quad (r^2 = 0{,}95) \qquad (11)$$

Auf diese Weise wurde die Konsumtion für die Mai- und Augustproben bestimmt.

Ergebnisse der qualitativen Nahrungsuntersuchungen

Die Untersuchungen zeigen, daß das Nahrungsspektrum des Barsches breit gefächert ist: neben planktischen Organismen (in der Hauptsache Cladoceren) ernährt er sich von benthisch lebenden Insektenlarven und zu einem geringen Teil auch von Kleinfischen.

Abgesehen vom unterschiedlichen Fraßverhalten einzelner Gruppen der Barschpopulation waren im Tagessowie im Jahreszyklus mehr oder weniger regelmäßig wiederkehrende Fraßgewohnheiten zu beobachten. Tageszeitliche Unterschiede in der Zusammensetzung der Nahrung sind ebenso wie Veränderungen in der Quantität der Konsumtion von der Aktivität der Fische bestimmt; Zooplankter werden beim Aufenthalt im Pelagial aufgenommen, am Grund Insektenlarven.

Bezüglich der Aufnahme an Zooplankton ist der Tagesgang der Konsumtion im Juli in der Tendenz zweigipfelig. Besonders die abendliche Pelagialwanderung ist mit einer höheren Freßaktivität gegenüber Zooplanktern verbunden (Abb. 1). Untersuchungen an kleinen Seen in Oberbayern (BOHL 1980) und am Hafnersee (SCHIEMER 1982) zeigen, daß es besonders nachts zu einer starken räumlichen Überlagerung von herbivoren Zooplanktern und zooplanktivoren Räubern kommt. Obwohl unsere Untersuchungen nur zwischen 5 und 21 Uhr durchgeführt wurden, scheint der Konsum, besonders von Zooplankton, in den Nachtstunden nicht höher als am Tage zu sein, eher scheint er in der zweiten Nachthälfte wesentlich niedriger als am Tage zu liegen. Hierauf weisen die sehr geringen Mageninhalte in den frühen Morgenstunden und der nicht sehr hohe Anstieg der Zooplanktonkonsumtion in den späten Abendstunden hin.

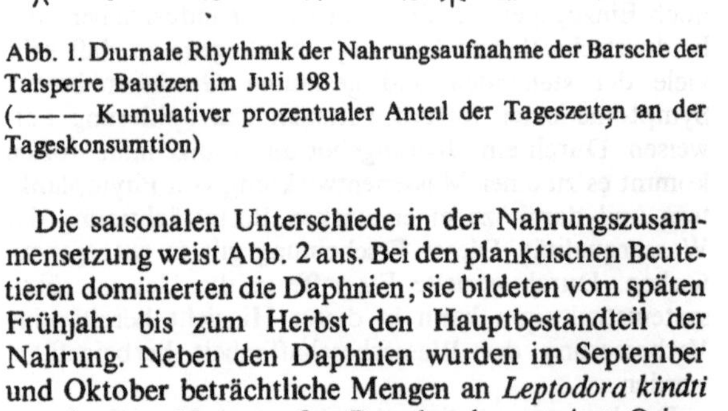

Abb. 1. Diurnale Rhythmik der Nahrungsaufnahme der Barsche der Talsperre Bautzen im Juli 1981
(——— Kumulativer prozentualer Anteil der Tageszeiten an der Tageskonsumtion)

Die saisonalen Unterschiede in der Nahrungszusammensetzung weist Abb. 2 aus. Bei den planktischen Beutetieren dominierten die Daphnien; sie bildeten vom späten Frühjahr bis zum Herbst den Hauptbestandteil der Nahrung. Neben den Daphnien wurden im September und Oktober beträchtliche Mengen an *Leptodora kindti* von den 15—20 cm großen Barschen konsumiert. Orientierende Untersuchungen an Barschen der 0-Gruppe im Litoral weisen auch im Juli und August eine erhöhte Aufnahme von *Leptodora* aus (Tab. 1). Copepoden, die im Plankton in hohen Individuenzahlen auftraten, waren nur im März und April, wenn noch keine starke Daphnienentwicklung zu verzeichnen ist, in nennenswerten Mengen in den Fischmägen zu finden. Benthische Organismen (v. a. Ephemeropteren- und Chironomidenlarven), werden im gesamten Zeitraum gefressen; besonders jedoch im Frühjahr (März—Mai) sowie im Frühsommer.

Hohe Anteile eines Nahrungstieres an der Konsumtion gehen einher mit hohen Abundanzen dieser Organismen und/oder dem Fehlen anderer, sonst bevorzugter

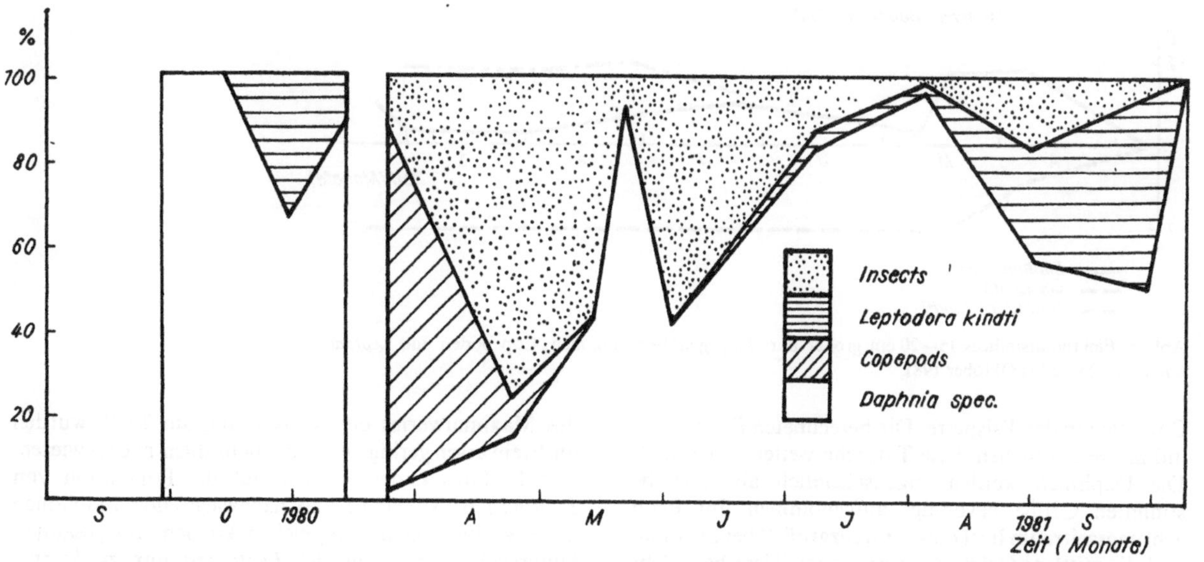

Abb. 2. Zusammensetzung der Nahrung von 15—20 cm großen Barschen in der Talsperre Bautzen vom Herbst 1980 bis Herbst 1981

Tab. 1. Nahrungszusammensetzung (% der Gesamtfeuchtmasse) des Barsches der Talsperre Bautzen im Langenbereich von 5—15 cm im Jahre 1982

Datum	n	Längenklasse (cm)	Cl	L	CH	MK	Fisch
29 7.	30	5—7	83	17	—	—	—
1. 7.	15	10—15	43	—	57	—	—
29. 7.	40	10—15	9	11	3	3	74
12. 8.	4	10—15	1	—	—	—	99
2. 11.	16	12—14	100	—	—	—	—

(Cl — Cladoceren, L — *Leptodora kindti*, CH — Chironomidenlarven, MK — Muschelkrebse)

Nahrungstypen. Eine bestimmte Tagesration ist also das Ergebnis
— der Abundanz der Nährtiere sowie
— des Auswahlvermögens der Fische.

Mittels des Elektivitätsindex (*E*) (IVLEV 1961) können potentielle Nahrungskomponenten hinsichtlich ihrer Bevorzugung in der Nahrung verglichen werden:

$$E = \frac{r_i - p_i}{r_i + p_i} \qquad (12)$$

r — Anteil einer Nahrungskomponente an der Gesamtkonsumtion (%);
p — Anteil dieser Komponente an der Nahrungskonzentration im Gewässer (%).

Um seine Beute lokalisieren zu können, muß sie für den Barsch visuell wahrnehmbar sein; chemische oder mechanische Perzeption spielt eine untergeordnete Rolle (VINYARD & O'BRIAN 1975, O'BRIAN 1979, ENDERLEIN 1981). Unterschiedlich für verschiedene Arten von Fischen kann eine „reaktive Distanz" definiert werden, als „Minimum der Distanz, bei welcher der Räuber seine Beute lokalisiert" (O'BRYIAN 1979). Die Größe dieser reaktiven Distanz ist in erster Linie abhängig von der Nahrungsgröße sowie der Färbung und Pigmentierung der Beute.

Eine solche Größenselektion wird bereits beim Betrachten der Struktur der Zooplanktongesellschaft deutlich: die untersuchte Längengruppe der Barsche der Talsperre Bautzen frißt selektiv die größeren Formen der im Gewässer vorhandenen Cladoceren (*Leptodora* sei aus der Betrachtung zunächst ausgeschlossen) (Abb. 3). Die mittlere Größe der in den Barschmägen gefundenen Daphnien liegt deutlich über der mittleren Größe der

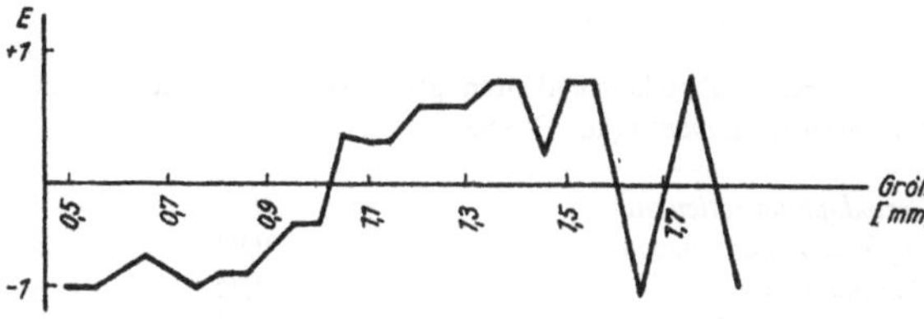

Abb. 4 Elektivitätsindices (E) der 15—20 cm großen Barsche für alle Größenklassen der Daphnien in der Talsperre Bautzen

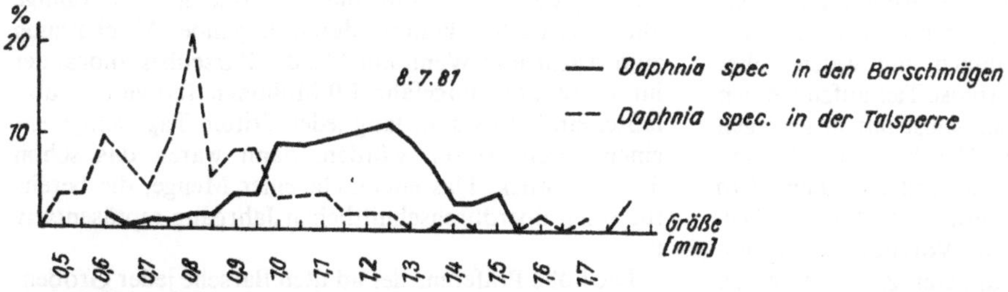

Abb. 3 Prozentuale Anteile der einzelnen Größenklassen der Daphnien in den Barschmägen sowie in der Talsperre
[Größenangaben für Daphnien im Gewässer aus KOSSATZ (1982)]

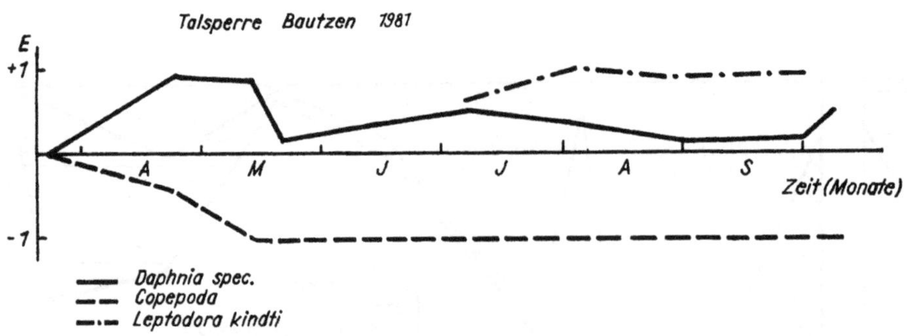

Abb. 5. Elektivitätsindices 15—20 cm großer Barsche gegenüber Daphnien, Copepoden und *Leptodora kindti* von März bis Oktober 1981

Daphnien in der Talsperre. Die berechneten Elektivitätsindizes verdeutlichen diese Tatsache weiterhin (Abb. 4). Die Daphnien werden augenscheinlich ab einer bestimmten Größe bevorzugt aufgenommen. Im Beobachtungszeitraum liegt diese „Grenzgröße" bei ca. 1 mm, und die mittlere Größe der gefressenen Tiere bei 1,1 bis 1,7 mm.

Die nur sehr geringe Konsumtion an Copepoden läßt sich nicht aus einem geringen Vorkommen dieser Tiere im Gewässer erklären, sie ist eher negativ korreliert mit dem Angebot an Daphnien. Auch über eine Größenselektion ist dieses Phänomen nicht zu deuten: die Daphnien sind nur wenig größer als die Cyclopoiden und Diaptomiden und müßten so von den Barschen gleichermaßen erkannt werden. Die Ursache dürfte allein in der Art sowie Geschwindigkeit der Fortbewegung der Beutetiere zu suchen sein (Tab. 2 und 3). Der Barsch als gut sehender Fisch ist also in der Lage, Informationen über das Fluchtverhalten verschiedener Nahrungstypen mit dem visuellen Erscheinen dieser Beutetiere zu korrelieren und in eine verschiedene Selektivität zu übertragen (IVLEV 1961, VINYARD 1980).

Tab. 2. Schwimmgeschwindigkeit von Zooplanktern (aus DRENNER et al 1978)

Daphnien	6 cm/s
Cyclops scutifer	30—40 cm/s
Diaptomus kenai	145 cm/s

Tab. 3. Fangwahrscheinlichkeiten gegenüber verschiedenen Zooplanktern für *Lepomis* (aus DRENNER et al. 1978)

Ceriodaphnia reticulata	0,96
Daphnia galeata mendotae	0,92
Daphnia pulex	0,76
Cyclopoide Copepoden	0,28
Cyclops scutifer	0,24
Diaptomus pallidus	0,07

Leptodora stellt für die Barsche auf Grund ihrer Größe ein Kettenglied beim Übergang zur piscivoren Lebensweise dar. Es ist jedoch erstaunlich, mit welch hoher Selektivität die Barsche dieses farblose Tier aufzunehmen vermögen. So ist es im Gewässer selbst nur in geringen Individuenzahlen zu finden. Im Vergleich zu allen anderen planktischen Nahrungstieren wurden für *Leptodora* die höchsten Elektivitätsindizes ermittelt (Abb. 5). *Leptodora* scheint für den Barsch die Vorzugsnahrung unabhängig von der Abundanz anderer Zooplankter zu sein. Im Beobachtungszeitraum stimmen jedoch die höchsten Konsumtionen an *Leptodora* nicht mit deren höchsten Abundanzen im Gewässer überein. Während des Maximalwertes der Konsumtion am 29. 9. wurden im Freiwasser geringste Individuenzahlen nachgewiesen. Der Fraßdruck der Barsche auf die Population von *L. kindti* ist so hoch, daß nachwachsende Individuen diese Verluste nicht ausgleichen können. Im gesamten Untersuchungszeitraum ist *Leptodora* nur zu kurzen Massenentwicklungen fähig, wenn der Fraßdruck der Fische infolge extrem hoher pH-Werte und/oder starker Trübung vorübergehend ausfällt (BENNDORF et al. 1988).

Beim Vergleich der einzelnen Nährtiere hinsichtlich ihrer Anteile am gesamten Nahrungsspektrum waren die Jungfische als potentielle Beute der Barsche überraschend schwach vertreten. Nur in den wenigsten Fällen waren in den Barschmägen noch wägbare Reste von Kleinfisch vorhanden, meist wiesen nur einige Ctenoidschuppen auf eine piscivore Lebensweise der Tiere hin. So blieben meßbare Anteile von Kleinfisch am Gesamtkonsum der Barsche unbedeutend (vgl. Abb. 2). Bezieht man aber jene Tiere in die Betrachtung mit ein, die Schuppen im Magen hatten, so lebten bis zu 12% der Barsche der untersuchten Längengruppen räuberisch (Abb. 6). Die

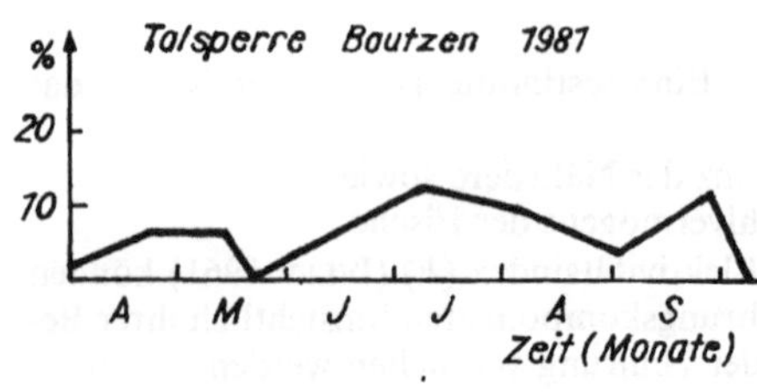

Abb. 6. Anteil der im Jahre 1981 untersuchten Barsche mit Fischresten im Magen

höchsten Anteile räuberisch lebender Barsche stimmen dabei mit höchsten Abundanzen an Kleinfisch im Juni und Juli überein. Nahrungsanalysen an 10—15 cm großen Barschen im Juli und August im Ufersaum weisen allerdings einen hohen Fischanteil in der Nahrung auf (bis nahezu 100%; vgl. Tab. 1). Die insgesamt große Bedeutung des Barsches für die Kontrolle der 0-Jahrgänge aller Fischarten kommt durch folgende Abschätzung zum Ausdruck: Wenn nur 5% des Barschbestandes, der im Jahre 1981 ungefähr 1,9 Millionen Individuen umfaßte, ein halbes Jahr lang jeden dritten Tag wenigstens einen Fisch fressen würden, dann wären das schon 5,7 Mio Stück. Das entspricht einer Menge, die bereits für einen überdurchschnittlichen Jahrgang repräsentativ ist.

Die hohe Präferenz der adulten Barsche jeder Größengruppe für Jungfische unter 10 cm Länge und das Hauptverbreitungsgebiet der Jungfische im Sommer in Ufernähe einerseits sowie die je nach Sichtverhältnissen stattfindende „Schutzwanderung" eines erheblichen Teils der adulten Barschpopulation bei Tagesanbruch aus der Ufernähe in tiefere Bezirke andererseits (THORPE 1977, SCHULTZ 1983), aber auch die relativ hohe Zahl von Barschen, die 1981 noch in der Talsperre vorhanden war, erklären den trotz hoher Präferenz niedrigen Anteil des Fisches in der Nahrung des Barsches in Längengruppen, die allgemein als piscivor gelten.

Nach SCHÖNERS Theorie (zit. in VINYARD 1980) selektiert ein Räuber diejenige Beute, die die größte Nettorate des Umsatzes erbringt. Beim Übergang zur piscivoren Lebensweise müssen für den Räuber die höheren energetischen Verluste beim Fang und die geringere Fangwahrscheinlichkeit durch den höheren Energiegehalt der Nahrung ausgeglichen werden. In welche Richtung diese Aufwandsoptimierung geht, hängt vom Verhältnis der Abundanzen an Kleinfisch und anderen Beutetieren einerseits und an Raubfischen andererseits sowie von einer Reihe abiotischer Faktoren ab, z. B. Licht und Wasserturbulenz. Der Übergang zur piscivoren Lebensweise kann damit recht variabel sein.

Für den Untersuchungszeitraum 1981/82 kann verallgemeinernd eingeschätzt werden:
Obwohl für die meisten der untersuchten Barsche die Daphnien noch den Hauptbestandteil der Nahrung ausmachen, nehmen die Barsche, sobald größere Beutetiere auftreten, diese bevorzugt auf. (Im Juni werden hauptsächlich Insektenlarven sowie von August—Oktober *L. kindti* gefressen.) Bei hohen Abundanzen an Kleinfisch im Gewässer sind die Anteile räuberisch lebender Barsche am höchsten.

Ergebnisse der quantitativen Untersuchungen

Der im Jahre 1981 ermittelte Konsum des Barsches in den Längengruppen von 15—21 cm an *Daphnia galeata* und *Leptodora kindti* ist in Tab. 4 aufgeführt. Es wurde versucht, die Höhe der Konsumtion (C) in Abhängigkeit von der Höhe des Nahrungsangebotes zu beschreiben. Man erhält dabei generell eine Sättigungskurve:

$$C = C_{max} \cdot \frac{Z}{K_f + Z} \tag{13}$$

C_{max} — die maximale Konsumtion an Zooplanktern;

K_f — die Biomasse der Zooplankter, bei der $C = \frac{1}{2} C_{max}$, (14)

$\dot{Z}$ — Biomasse an Zooplanktern;
C — Tageskonsumtion an Zooplanktern.

Hinsichtlich der Daphnien als Nährtiere wurde für die Barsche der Talsperre Bautzen (Länge: 15—21 cm) die in Abb. 7 dargestellte Abhängigkeit ermittelt. Aus dem in Abb. 7 gezeigten Kurvenverlauf und der von KOSSATZ (1982) im Untersuchungszeitraum Mai—Oktober 1981 gefundenen maximalen Daphnienbiomasse von 12,56 mg FM $\cdot$ l^{-1} resultiert eine maximale Konsumtion von 1246 mg FM $\cdot$ d^{-1} $\cdot$ Fisch^{-1}. Dieser Wert entspricht bei Annahme einer individuellen Fischbiomasse von 62 g (zur Zeit des maximalen Daphnienangebots) einer Konsumtionsrate von 2,0% bzw. bei einer Stückmasse von 78 g (Mittelwert des Jahres 1981) einer Rate von 1,6%.

Die in der berücksichtigten Literatur (SWENSON & SMITH 1973, POPOVA & SYTINA 1977, SWENSON 1977, THORPE 1977, ELLIOT & PERSSON 1978, COSTA 1979, NYBERG 1979) allerdings nach unterschiedlichen Verfahren angegebenen Werte für den Tageskonsum von Barschartigen (*Perca fluviatilis, P. flavescens, Stizostedion vitreum vitreum*) zeigen eine beträchtliche Schwankungsbreite von 0,1—6,5% des Frischgewichtes der Fische. Die für die Barsche der Talsperre Bautzen ermittelten Konsumtionsraten sind im unteren Drittel dieser Spanne einzuordnen. Die Konsumtionsraten für Daphnien und *Leptodora* bringen zusammen im Mittel der einzelnen Probenahmen zwischen 0,4 und 2,0%, wobei beide Extremwerte im Mai registriert wurden. Konsumtionsraten für einzelne Längenklassen sind für einige Probenahmetermine in Tab. 5 zusammengestellt. Die genannten Werte sind sicherlich um einen gewissen Betrag unterschätzt, weil (a) Probenahmen in den Nachtstunden fehlten und (b) auf Grund der realisierten Probenahmeabstände nur ein vereinfachendes Modell zur Berechnung angewandt werden konnte. Durch beide Fehlerquellen mag eine Unterschätzung um etwa 40 bis möglicherweise 80% eingetreten sein (vgl. auch maximale Konsumtionsrate bezüglich Daphnien: 1,6% bzw. 2,0%).

Bei der Ermittlung der Gesamtkonsumtionsrate ist der Konsum an anderen Nahrungskomponenten (Insektenlarven, Fische) hinzuzurechnen. Der wirkliche Anteil räuberisch lebender Barsche scheint nur wenig unter den geschätzten Werten zu liegen, wenn in den Sommermonaten leere Mägen nicht als Indikator für Fischfraß berücksichtigt wurden, da leere Mägen im Sommer nur in Einzelfällen auftraten. LESMARK (1983) gibt für zwei schwedische Seen (Läen-See und Björnsjön-See) ähnliche Ergebnisse an, wie wir sie für die Talsperre Bautzen ermittelten. Auch dort ist die Konsumtionsrate von Barschen in den Längengruppen 15—19 cm sehr niedrig, sie stieg erst mit der Aufnahme von Fischnahrung bei Barschen der Längenklassen ≧20 cm beträchtlich an.

Aus den dargestellten Befunden wird deutlich, daß der Fraßdruck des Barsches, der in den Jahren 1981/82 75% des planktonfressenden Fischbestandes in der Talsperre Bautzen ausmachte, auf das Zooplankton nicht hoch war. Nach Berechnungen von SCHULTZ (1983)

Tab. 4. Tageskonsumtion des Barsches bezüglich *Daphnia galeata* und *Leptodora kindti* in der Talsperre Bautzen im Jahre 1981

Datum	Anzahl der Probenahmen	Anzahl der Verdauungsversuche	Anzahl der untersuchten Barsche	Konsumtion an *D. galeata* (mg/d/Fisch)	Konsumtion an *L. kindti* (mg/d/Fisch)
12.—14. 5.	4	1	76	171,5	0
21. 5.	2	1	89	1288,1	0
1.—2. 6.	7	4	173	564,0	0
7.—8. 7.	6	6	250	563,4	21,1
4. 8.	2	—	54	1015,5	0
31. 8.—1. 9.	5	2	109	263,5	133,9
28.—29.9/8. 10.	4	4	144	466,9	628,0

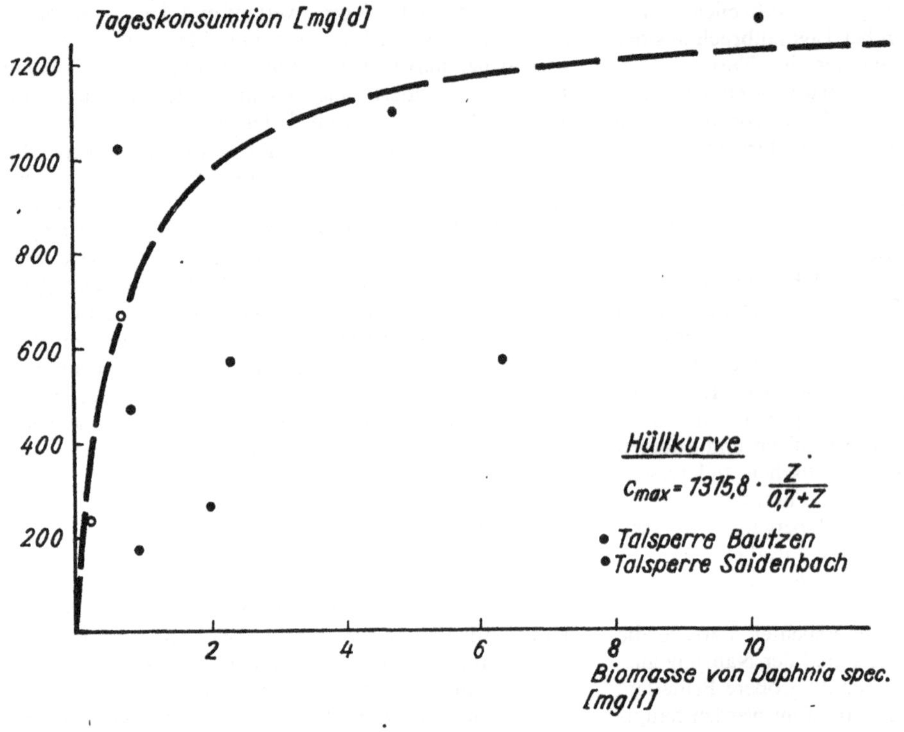

Abb. 7. Beziehung zwischen der Konsumtion des Barsches an Daphnien und dem Nahrungsangebot an Daphnien im Gewässer
[Biomassedaten der Daphnien in der Talsperre Bautzen nach KOSSATZ (1982), in der Talsperre Saidenbach nach W. HORN (pers. Mitt.)]

Tab. 5. Konsumtionsraten der Barsche verschiedener Längengruppen bezüglich *Daphnia galeata* und *Leptodora kindti*

Datum	Längengruppe (cm)	Barschgewicht (g)	Tageskonsumtionsrate (%) *D. galeata*	*L. kindti*	Insgesamt
1./2. 6. 81	15	46,7	1,21	—	1,21
	16	57,1	0,99	—	0,99
	17	64,4	0,88	—	0,88
	18	84,6	0,67	—	0,67
7./8. 7. 81	16	60,0	0,94	0,03	0,80
	17	68,4	0,82		
	18	80,4	0,70		
	19	93,2	0,60		
1. 9. 81	16	64,3	0,41	0,21	0,62
	17	76,8	0,34	0,17	0,51
	18	88,2	0,30	0,15	0,45
	19	101,5	0,26	0,13	0,39
	20	112,6	0,23	0,12	0,35
	21	128,0	0,21	0,10	0,31

ging der Fraßdruck der Barsche auf die Daphnien im Jahre 1981 um reichlich die Hälfte gegenüber 1978/79 zurück.

Zusammenfassung

Als Bestandteil eines Langzeitexperimentes zur Biomanipulation in der hypertrophen Talsperre Bautzen wurden im Zeitraum von 1980 bis 1982 qualitative und im Jahre 1981 quantitative Nahrungsuntersuchungen an Barschen durchgeführt. Insgesamt wurden 1231 Barsche im Längenbereich von 15—20 cm und 105 Tiere im Längenbereich von 5—15 cm bearbeitet.

Die Tageskonsumtion wurde in Freilanduntersuchungen und Verdauungsversuchen überwiegend nach einer Methode von THORPE (1977) bestimmt.

Trotz diurnaler und saisonaler Unterschiede in der Nahrungszusammensetzung fraßen Barsche von etwa 10 cm Größe an bevorzugt große Beuteobjekte wie Fische, Insektenlarven und *Leptodora kindti*. Eine Selektion der Beute ist ebenso bezüglich der Daphnien festzustellen: 15—20 cm große Barsche fraßen nur Daphnien in einer Größe über 1,0 mm

Die Abhängigkeit des Tageskonsums an Daphnien vom Nahrungsangebot wird näherungsweise mit einer Hyperbelfunktion beschrieben. Der Tageskonsum an Daphnien und *Leptodora kindti* wurde im Zeitraum von Mai bis Oktober bei Barschen von 15—20 cm Länge und 40—100 g Stückmasse mit 0,2—1,3 g je Fisch berechnet. Wenn auch die Konsumtion mit diesen Werten um einen Betrag von etwa 40 bis möglicherweise 80 % unterbewertet ist, so war dennoch der Fraßdruck der Barsche auf die Daphnienpopulation nach überschläglichen Berechnungen um reichlich 50 % im Vergleich zu den Jahren vor der Biomanipulation (1978 und 1979) herabgesetzt.

Die Barschpopulation des Jahres 1981 vernichtete Fische (hauptsächlich Perciden) in einer Menge, die der Anzahl an Jungfischen eines überdurchschnittlichen Barschjahrgangs in der Talsperre entspricht. Dem Gesamtkonsum ist ferner eine gewisse Menge an Insektenlarven hinzuzurechnen.

References

BENNDORF, J., SCHULTZ, H., BENNDORF, A., UNGER, R., PENZ, E., KNESCHKE, H. KOSSATZ, K., DUMKE, R., HORNIG, U., KRUSPE, R., & REICHEL, S (1988): Food-web manipulation by enhancement of piscivorous fish stocks · Long-term effects in the hypertrophic Bautzen reservoir *Limnologica* (Berlin) **19** (1), 97—110

BOHL, E. (1980): Diel pattern of pelagic distribution and feeding in planktivorous fish. Oecologica (Berlin) **44**, 368—375.

COSTA, H. H (1979) The food and feeding chronology of yellowperch in Lake Washington. Int. Rev. ges. Hydrobiol. **64** (6), 783—793.

DRENNER, R. W., STRICKLER, J. R, & O'BRIAN, W. J (1978). Capture probability: The role of zooplankter escape in the selective feeding of planktivorous fish. J. Fish. Res Bd Can. **35** (10), 1370—1373.

ELLIOT, J. M. (1972)· Rate of gastric evacuation in brown trout (*Salmo trutta*). Freshwater Biol. **2** (1), 1—18

— & PERSSON, L (1978)· The estimation of daily rates of food consumption for fish. J Anim. Ecol **47**, 977—991.

ENDERLEIN, O. (1981)· When, where, what and how much does the adult cisco, *Coregonus albula* (L.) eat in the Bothnian Bay during the ice-free season. Inst. Freshwater Res, Drottningholm, Rep. Nr. **59**, 21—32.

IVLEV, V S. (1961): Experimentalnaja ekologija pitanija ryb Kiew, 271 S.

KOSSATZ, K. (1982)· Struktur und Dynamik wichtiger Zooplankton-Populationen in der Talsperre Bautzen und im Restgewässer Gräfenhain. Diplomarb., Techn. Univ Dresden, Fachrichtung Techn. Hydrobiol, 1—57.

LESMARK, O (1983): Influence of abiotic and biotic factors on the structure of perch and roach populations in thirteen Swedish lakes, with special reference to interspecific competition. In. O. LESMARK, Competition between perch (*Perca fluviatilis*) and roach (*Rutilus rutilus*) in South Swedish Lakes. Diss., Lund, 135—172.

NYBERG, P. (1979): Production and food consumption of perch, *Perca fluviatilis* L., in two swedish forest lakes. Inst. Freshwater Res., Drottningholm, Rep. Nr. **58**, 140—157.

O'BRIAN, W J. (1979): The predator — prey interactions of planktivorous fish and zooplankton Amer Sci. **67** (5), 572—581

POPOVA, O. A., & SYTINA, C. A. (1977)· Food and feeding relations of Eurasian perch (*Perca fluviatilis*) and pike-perch (*Stizostedion lucioperca*) in various waters of the USSR. J. Fish. Res. Bd. Can. **34** (10), 1559—1570

SCHIEMER, F. (1982): Trophische Interaktionen im Pelagial des Hafnersees, Kärnten. Sitzungsber Österr. Akad Wiss., Mathem. naturwiss. Klasse, Abt. I, **191** (5—10), 209—230.

SCHULTZ, H. (1983): Untersuchungen über die Größe und Struktur des Fischbestandes sowie den Nahrungskonsum ausgewählter Fischarten in zwei Stauseen unterschiedlichen Trophiegrades. Diss. A, Techn. Univ. Dresden, Fak. f. Bau-, Wasser- und Forstwesen, 1—141.

SWENSON, W. A. (1977): Food consumption of walleye (*Stizostedion vitreum vitreum*) and sauger (*S. canadense*) in relations to food availability and physical conditions in Lake of the Woods, Minnesota, Shagawa Lake and Western Lake Superior. J. Fish. Res. Bd. Can. **34**, 1634—1654.

— & SMITH, L. L. (1973): Gastric digestion, food consumption, feeding periodicity, and food conversion efficiency in walleye (*Stizostedion vitreum vitreum*) J. Fish. Res. Bd. Can **30** (9), 1327—1336.

THORPE, J E. (1977): Daily ration of adult perch (*Perca fluviatilis*) during summer in Loch Leven, Scotland J. Fish Biol. **11** (1), 55—68.

VINYARD, G. L. (1980): Differential prey vulnerability and predator selectivity · Effects of evasive prey on bluegill (*Lepomis machrochirus*) and pumpkinseed (*L gibbosus*) predation Can J. Fish Aquat. Sci. **37** (12), 2294—2299.

— & O'BRIAN, W. J (1975): Dorsal light response as an index of prey preferences in bluegill (*Lepomis machrochirus*). J. Fish. Res Bd. Can. **32** (10), 1860—1863.

WINDELL, J T. (1970)· Food analysis and rate of digestion In · W E. RICKER (ed.), Methods for assessment of fish production in freshwaters IBP-Handbook No. 3, 2nd .ed., 197—203. Oxford and Edinburgh,

Eingegangen am 2 Februar 1987

Anschrift der verantwortlichen Autoren:

Dipl.-Biol. UTE KÖPKE
Dr. HEINZ SCHULTZ
Technische Universität Dresden
Sektion Wasserwesen
Bereich Hydrobiologie
Mommsenstr. 13
Dresden
DDR-8027

Limnologica (Berlin) 19 (1), 45—49, Juli 1988

Department of Fish Culture and Fisheries, Wageningen, The Netherlands; Limnological Institute, Oosterzee, The Netherlands; Organization for the Improvement of Inland Fisheries (OVB), Nieuwegein, The Netherlands

Possibilities for Stock Enhancement of Pikeperch (*Stizostedion lucioperca*) in Order to Increase Predation on Planktivores

By W. L. T. VAN DENSEN & M. P. GRIMM

With 3 Figures

Key words: *Stizostedion*; Stocking; Biomanipulation

Abstract

Regulation of fishery for pikeperch and stocking with 0- and 1-group pikeperch are discussed as possible strategies of pikeperch stock enhancement to increase the predation on planktivores. In early summer stocking with 0-group or 1-group pikeperch of a large individual size relative to the size of the naturally produced pikeperch, seems to be a promising technique. In these situations, availability of prey fish is large and this positively influences growth and survival of the individuals stocked. Supplementary fishing for large planktivores such as adult bream might be required.

Introduction

Pikeperch is the main predatory fish in the open water zone of most eutrophic water bodies in Central and Northern Europe. Because of the increasing eutrophication in the last half century, the position of pikeperch has become even more dominant. The amount of pike, related to the surface of the vegetated area (GRIMM 1983), markedly declined, and also in smaller water bodies, pikeperch could become the dominant predatory fish. Further enhancement of pikeperch stocks could modify the structure of the food web and thus, indirectly the position of herbivorous zooplankton such as *Daphnia*. A strong impact of pikeperch on the populations of small planktivorous fish causes a less intensive, size-selective predation by these planktivores on the large herbivorous zooplankton. Planktivore-induced changes in algal biomass were often described (LEAH et al. 1980, HESSEN 1983, SHAPIRO & WRIGHT 1984, ANDERSSON 1985). In this way, enhancement of *Stizostedion* populations could be used as a tool in biomanipulation experiments directed to an overall enlargement of the grazing pressure on the populations of algae (SHAPIRO & WRIGHT 1984, BENNDORF & HORN 1985).

As in every fish population, the stock of pikeperch is determined by recruitment (egg, larval, one year-old), growth, natural and fishery mortality. Insight in these processes forms the basis of strategies for pikeperch stock enhancement. Together with information on the foraging behaviour, the extent of the influence of an enlarged stock of predatory fish on prey populations, especially those of planktivorous fish, can be predicted.

Determining factors

Recruitment

Generally, pikeperch shows large variations in year-class strength. The ratio of numerical density between the strongest and the weakest year-class can be 15 (WILLEMSEN 1977, Table 5) to more than 100 (SVÄRDSON & MOLIN 1973, Table 5). These large variations result in variable annual catches in professional fishery. Stocking programmes for 0-group pike perch in the Netherlands were sometimes used to cope with these variations, aiming at a more stable yield. In a heavily exploited natural population of pikeperch in IJsselmeer, the Netherlands, no correlation was found between larval density and the density of one-summer old pikeperch in autumn (WILLEMSEN 1977). A significant correlation was found between the density of one-summer olds and the presence of the year-class in the yield of the professional gillnet fishery two or three years later. The variation in the number of one-summer old pikeperch in this exploited pikeperch population was largely explained by the variation in summer temperature. Both, the density and the mean individual size of the 0-group pikeperch in autumn, were positively correlated with summer temperature.

In another unexploited pikeperch population in Tjeukemeer, the Netherlands, strong year-classes were also characterized by a large mean individual size of the 0-group in autumn (VAN DENSEN 1985a). The abundance of the slender 0-group smelt, which is an ideal prey item for pikeperch from the beginning of the summer onward, was important for the formation of strong year-classes. The high abundance of smelt improved the survival of the 0-group of pikeperch in two ways. The fast growth of 0-group piscivores, based on the consumption of smelt, made them less vulnerable for piscivorous predators, including adult pikeperch. Second, the high smelt abundance lowered the chance of every single 0-group fish, including 0-group pikeperch, to be preyed. This buffering effect has already been described for 0-group walleye, *Stizostedion v. vitreum* (FORNEY 1976).

The amount of smelt in the IJsselmeer is more stable than that in Tjeukemeer (pers. comm. J. WILLEMSEN). Probably for this reason, the year-class strength of pikeperch in IJsselmeer is determined more by variations in temperature.

Growth

Strong year-classes of pikeperch contain individuals of a mean size of about 15 cm and larger at the end of the first growing season. Mean growth from this size onward is mainly temperature-limited or limited by the duration of the growing season in Northern Europe. In IJsselmeer (52—53′ N), mean sizes per year-class are of a total length of 15—30—42—50—57—63 cm (WILLEMSEN 1977).

Natural mortality .

Intra-specific predation seems to be the main reason for death of the 0-group and partially of the 1-group pikeperch. Adult pikeperch can prey on 1-group congeners. It is improbable that piscivores of a smaller gape width such as perch and eel will take pikeperch larger than 10 cm. Adult mortality is low and might be brought into relation with energy depletion after spawning, because in early summer dead specimens are sometimes found in the littoral zone.

Fishery mortality

In most countries, the minimum size allowed to be landed is of 40—45 cm length.

Smaller pikeperch might endure high mortality as by-catch in the fykenet fishery for eel. The minimum size roughly corresponds with the optimum size of pikeperch caught in gillnets of about 90 mm (stretched mesh). In the IJsselmeer the total mortality of pikeperch in the exploited population amounted to 81%. This corresponded to an instantaneous mortality rate of 1.66 per year (Willemsen 1977).

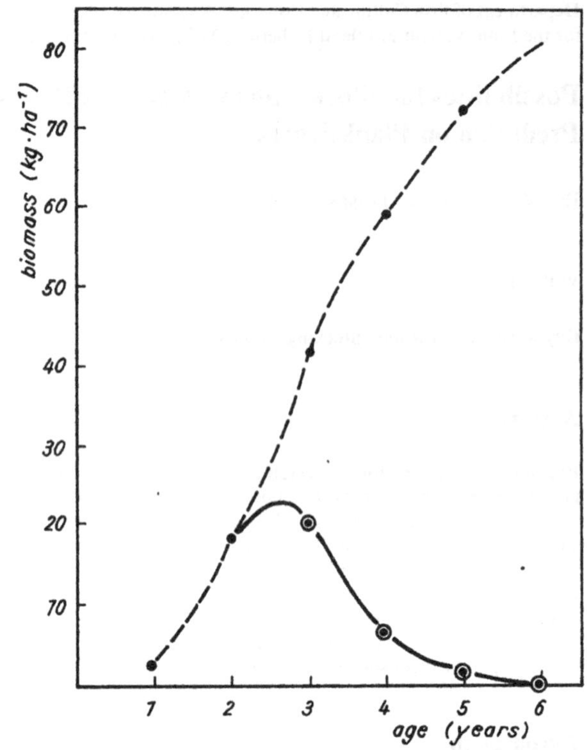

Fig. 1. Biomass of a pikeperch year-class with and without a commercial gillnet fishery. Based on figures in Willemsen (1977) and using a LW-relationship·W = 0.0058 $L^{3.1}$ (W in g; L in cm total length). The pikeperch recruits to the fishery at the age of 2.5 years

Stock enhancement

Regulation of fishery

A first stage in enlarging the stock size of pikeperch is to strictly regulate professional and recreational fishery. When this is done, the pikeperch population structure will mature because the rejuvenation process as a consequence of fishing-out the larger individuals will come to a stand-still. When only one strong year-class dominates the population before the measurement was taken, the growth in total biomass will nearly follow the pattern of this year-class. By simply accounting for an instantaneous mortality rate of 0.2 per year, the growth in biomass in a situation of an abundant food supply, like in IJsselmeer, will follow a pattern shown in Fig. 1 in the first period.

The constant fishing mortality assumed can only be an approximation. In reality, there is a complex size-selective mortality depending on growth rate, variance in the size distribution of the year-class and the selectivity of the gillnets per size class of the pikeperch.

When two strong year-classes succeed one another there will be no influence on survival as long as the two year-classes can only interfere via the common food resource. Once two years old, it seems improbable that the second year-class will experience intra-specific predation (cannibalism). Cannibalism within the year-class after the first summer is also most improbable, because the coefficient of variation in the length-frequency distribution is low (5—7% for two years and older pikeperch). By closely following growth and condition of the adult pikeperch, the stage of possible food limitation and density-dependent growth can be detected.

Together with the occurrence of food-limited growth, the survival of prey fish, including 0-group pikeperch, will be depressed. The recruitment of the new year-classes can be blocked-off by intra-specific predation for several years, and the maturation of the pikeperch population structure will continue. Pikeperch biomass might be already food-limited when recruitment is depressed by intra-specific predation. But also the opposite situation may occur. In the latter case, the prey populations will not be used sufficiently and only stocking of pikeperch of low vulnerability to predation could further enhance pikeperch stocks.

Stocking

Pikeperch used for stocking purposes in the Netherlands, generally are pond-raised one-summer old individuals of 8—12 cm. For the introduction of pikeperch, these individuals proved to be of use. In the new environment, they are fit to survive their first winter. However, if stocked to enhance pikeperch populations already present, recruitment to the adult stock seems low (Grimm 1982). Their survival appears to be the same as for other planktivorous 0-group pikeperch. Their contribution to 1-group and older pikeperch was low, probably due to intra-specific predation in the period from October to March (Grimm, unpubl. results). When 0-group pikeperch is stocked at the beginning of the summer, at a size relatively large in comparison to the 0-group pikeperch recruited naturally, in most cases they will meet a high density of prey fish available. The availablity of prey fish is a function of the species composition of the prey populations, of the prey size relative to the size of the pikeperch and of the prey size distribution.

The most slender prey species is the smelt, *Osmerus eperlanus*, but its distribution is confined to areas near the sea. Although the length of the smelt in the stomachs of 5—10 cm pikeperch can be more than 50% of the length of the pikeperch, under natural conditions only the smallest smelt is available for the largest 0-group pikeperch in mid-summer. These piscivorous pikeperch belong to the largest specimens in the positively skewed size distribution of 0-group pikeperch (Fig. 2). When the mean size of the pikeperch would be only a few centimetres larger, the availability of prey fish would strongly increase. This mechanism was demonstrated in drainable

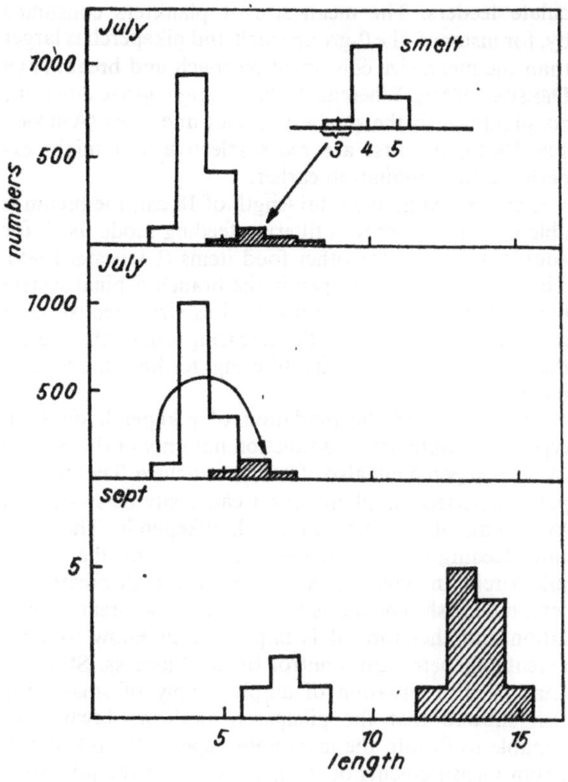

Fig. 2. The LF-distributions of 0-group pikeperch and potential prey fish. The shaded part of the size distribution indicates piscivorous 0-group pikeperch. Prey fish in July can be for instance smelt (upper) or 0-group pikerperch (cannibalism)

ponds containing a spawning population of pikeperch and cyprinids. The 0-group is capable to prey on its congeners from 10 mm onward and on cyprinid larvae from 15—20 mm onward. Here, two batches of 0-group pikeperch (3—5 cm) were stocked with a 3 week interval. This was done to simulate a proper tuning and a retardation of pikeperch length relative to that of prey fish (Fig. 3).

Generally, prey species occurring in high densities in eutrophic waters at the beginning of the summer are, beside 0-group pikeperch, perch, roach and bream. The 0-group perch is present in the open water zone, as pikeperch and smelt, from the beginning of the summer onward. The 0-group bream and roach migrate to the open water zone at a later stage, in midsummer. The availability of these species is correlated with their body depth-length ratio, which is most unprofitable for the depp-bodied bream. Pikeperch and roach are more available, although 0-group bream can be found in 0-group pikeperch stomachs when the mean size of 0-group bream is small. The latter happens in cold summers (VAN DENSEN 1985a).

Stocking with indoor-raised pikeperch at the beginning of the summer could ensure the development of a 0-group cohort of piscivorous pikeperch attaining a size well exceeding 15 cm at the end of the summer. The stocking densities have to be in accordance with a probable survival rate and even more so with a desired density at the end of the summer. Stocking densities cannot be too high because otherwise they would over-exploit the prey populations present and would manifest a food-limited growth switching over to the consumption of zooplankton. The 0-group of prey fish might then be eliminated, but all planktivores are more efficiently controlled when the pikeperch year-class has grown older and became larger.

In order to have the maximum effect of a stocked pikeperch year-class, it might be useful first to cull the pikeperch present to assure optimum conditions for the survival of stocked 0- or 1-group pikeperch.

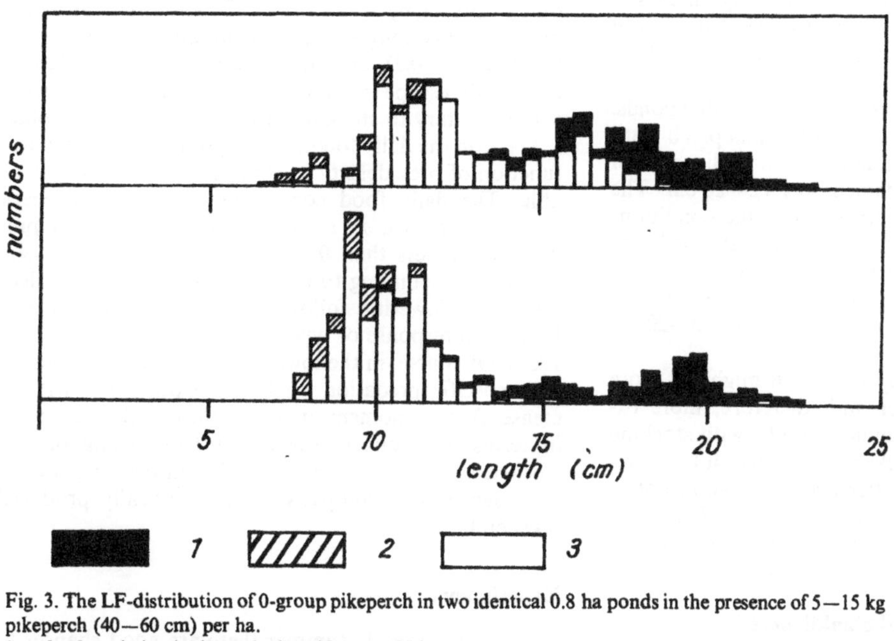

Fig. 3. The LF-distribution of 0-group pikeperch in two identical 0.8 ha ponds in the presence of 5—15 kg pikeperch (40—60 cm) per ha.
1 = 3—5 cm individuals stocked on 22 June 1984;
2 = 3—5 cm individuals stocked on 25 July 1984;
3 = naturally recruited individuals.

Culture

Traditionally, the culture of pikeperch in the Netherlands was restricted to those of 0-group individuals (8—12 cm in autumn) in ponds. The production of these pikeperch started as a secondary activity in ponds managed as carp production units. In this situation, pikeperch feeding was confined to the pelagical zooplankton. Nowadays, ponds may be managed for the production of pikeperch only. But as production generally aims at the highest number per ha by suppression of intra-specific predation, only planktivorous individuals not exceeding a length of 12 cm are produced.

For the economical production of viable pikeperch longer than 12 cm, at the end of the summer, two strategies can be followed corresponding with different sizes at stocking. The first product is already produced on a large scale, the afore named 8-12 cm 0-group pikeperch. However, instead of stocking them at the end of the first growing season, they should be over-wintered in the rearing ponds and stocked as 1-group pikeperch in mid—July of the next summer.

The alternative is to stock 2—3 cm 0-group pikeperch in early summer. These individuals can be indoor-raised on a large scale. To that extent, the methods described by Tölg (1981), Steffens (1981) and Schlumpberger & Schmidt (1980) were adapted using very low light intensity for 24 h periods and a frequent grading to sort out the fastest growing (cannibalistic) individuals of 10 mm and of greater length (Grimm, in prep.). These pikeperch are generally 10 mm longer than those naturally produced at the time of stocking (end of May, beginning of June). This, at least theoretically, enables them to forage even on cyprinid larvae.

Survival and growth of pikeperch stocked

The OVB (Organization for the Improvement of Inland Fisheries) in the Netherlands has stocked fish ponds and natural waters with pikeperch, marked by fin amputation (Patrick & De Haas 1971), and calculated the survival rates. Growth and survial of the following three size categories were 1.5, 2.5 and 10 cm stocked at the end of May, beginning of June and at the beginning of July, respectively.

The 1.5 cm pikeperch was released in drainable ponds. Their survival in the absence of older pikeperch until autumn was 55%, corresponding to a mortality rate of 0.5% per day. The size attained was 12—20 cm. The release of 2.5 cm pikeperch under the same conditions resulted in a survival of 10%. The size attained was 12—25 cm. Varying results were obtained with stocking natural waters with 2.5 cm pikeperch from no survival at all until a survival rate of 10% and a size of 12—20 cm in autumn.

Stocking of larger 0-group pikeperch might be more expensive because of longer and, therefore, more expensive culturing. The preliminary results with stocking natural waters with 10 cm pikeperch in early July, however, were most promising. After two years, almost 50% was still surviving. Sizes attained were 21—27 cm after the first summer, and 35—40 cm after the second summer.

Impact on the populations of planktivores

The major planktivores in the eutrophic water bodies in Northern Europe can be grouped into two categories depending on their feeding mode: particulate feeders and (pump)filter-feeders. The 0-group of most fish species are particulate-feeding planktivores during the whole or a part of their first growing season. Smelt, perch, roach and bream can continue this feeding mode in their second year and even later. A specific distinction is to be made in size-selectivity within the group of particulate feeders. The mean size of plankters consumed by, for instance, the 0-group smelt and pikeperch is larger than the mean size consumed by roach and bream (van Densen 1985b). Whereas the latter might graze down the zooplankton to the smallest species and sizes (Andersson 1985), the first are more selective and might experience food limitation earlier.

Once exceeding the total length of 18 cm, the bream is able to switch over to a filtering feeding mode, both for zooplankton and for other food items (Lammens 1985). The narrow space left open in the branchiospinal system makes a 18—20 cm bream even less size-selective than its smaller congeners. With increasing size of the bream, the size selectivity in its filtering feeding mode also increases.

The impact of the predation of pikeperch on both types of planktivores is a function not only of the density of the pikeperch but also of its size structure. The smallest particulate-feeding planktivores can easily be swallowed by 30 cm, about two years old, pikeperch. The large filter-feeding bream can only be preyed by the largest pikeperch. In typically eutrophic pikeperch-bream waters, most fish biomass is locked-up in the bream population and therefore, it is important to know to what extent pikeperch can control bream biomass. Stomach analysis in a situation of ample supply of small prey fish showed that the pikeperch swallows bream just capable to fit into the maximum gape of the pikeperch. From measurements of the gape width of the pikeperch and of the body height of the bream, both as a function of their total lengths (cm), the following relationship was found:

$$\text{bream length} = 0.308 \text{ pikeperch length} + 1.3$$

This implies that a bream exeeding 18 cm in length can only be preyed by 54 cm and larger pikeperch. Generally, most breams longer than 20 cm and 3—5 years old (Cazemier 1982) and manifesting a filtering-feeding mode, will escape predation. For this category, trawling supplementary to stocking of pikeperch can be applied.

Data on the daily food consumption of pikeperch as a function of prey density in natural waters are not available. The daily food consumption of the congeneric walleye, *Stizostedion v. vitreum*, decreased when prey density was less than 0.4 g per cubic meter (Swenson 1977), corresponding to only 8 kg per ha in a 2 m deep lake. This lowering followed the pattern typical of a functional response of type II on prey density (Holling 1965). Of course, in the lower range of prey densities the intra-specific predation on 0-group pikeperch will increase. At this moment, it is difficult to tell what profit, in terms of stronger escapement from being preyed, stocked 2.5 and 10 cm pikeperch has in the range of low prey densities in comparison with naturally produced pikeperch.

Experiments

Technically, in early summer there are good chances to enhance pikeperch stocks by stocking with 0-group or 1-group pikeperch.

The choice for the economically most efficient stocking strategy depends, among others, on the costs of culturing different sizes of pikeperch and on their survival rates in the natural environment. As far as survival rates are concerned, these cannot be easily predicted by data on pikeperch abundance, population structure and availability of prey species. To develop a model for an optimal stocking strategy, more experiments are required, some of which can be performed in the controlled environment of fish ponds.

Experiments have to be done in natural waters, to figure out for what time a pikeperch stock is enhanced after regulation of the fishery, stocking of 0- or 1-group pikeperch and possible supplementary fishing large planktivores. Especially, the negative feed-back of an enhanced pikeperch stock on its own somatic growth and recruitment has to be investigated.

Such experiments in natural waters can also reveal to what extent the changes in the fish fauna, observed after the introduction of pikeperch in new environments, are a result of pikeperch predation or of concomittant changes in the habitat (Fickling & Lee 1983). Anyway, there are situations where the introduction of pikeperch or the termination of fishery of the congeneric walleye is feared because of their strong influence on the populations of prey fish (Linfield 1984, Knight et al. 1984).

Zusammenfassung

Es werden methodische Probleme beim Einsatz von Zandern (*Stizostedion lucioperca*) der Klassen 0 und 1 zur Reduktion planktivorer Fische in Seen behandelt. Die zusätzliche Abfischung großer planktivorer Fische (z. B. adulte Bleie) wird für notwendig erachtet.

References

Andersson, G. (1985): The influence of fish on eutrophic lake ecosystems. In: Andis, Lakes pollution and recovery, p. 50—53. Rome.

Benndorf, J., & Horn, W. (1985): Theoretical considerations on the relative importance of food limitation and predation in structuring zooplankton communities. Arch. Hydrobiol., Beih. Ergebn Limnol. **21**, 383—396.

Cazemier, W. G. (1982): The growth of bream (*Abramis brama*) in relation to habitat and population density. Hydrobiol. Bull. **16**, 269—277.

Densen, W. L. T. Van (1985a): Piscivory and the development of bimodality in the size distribution of 0+ pikeperch (*Stizostedion lucioperca* L.). Z angew. Ichthyol. **1**, 119—131.

— (1985b): Feeding behaviour of major 0+ fish species in a shallow eutrophic lake (Tjeukemeer, The Netherlands). Z. angew. Ichthyol. **1**, 49—70.

Fickling, N. J., & Lee, R. L. G. (1983): A review of the ecological impact of the introduction of the zander (*Stizostedion lucioperca* L.) into waters of the Eurasian mainland Fish. Mgmt. **14**, 151 to 155.

Forney, J. L. (1976): Year-class formation in the walleye, *Stizostedion vitreum vitreum*, population of Oneida Lake, New York, 1966—1973. J. Fish. Res. Bd. Can. **33**, 783—792.

Grimm, M. P. (1982): Annual report 1980—1981. Organization for the Improvement of Inland Fisheries. OVB, Nieuwegein, The Netherlands (in Dutch).

— (1983): Regulation of biomasses of small (41 cm) northern pike (*Esox lucius* L.) with special reference to the contribution of individuals stocked as fingerlings (4—6 cm). Fish. Mgmt. **14**, 115—135.

Hessen, D. O. (1983): Introduction of pike, *Esox lucius*, to a small pond; effects on planktivorous fish, zooplankton and phytoplankton. Fauna **36**, 119—124.

Holling, C. S. (1965): Functional response of predators to prey density and its role in mimicry and population regulation. Mem. Entomol. Soc. Can. No. **45**.

Knight, R. L., Margraf, F. J. & Carline, F. (1984): Piscivory by walleyes and yellow perch in Western Lake Erie. Trans. Am. Fish. Soc. **113**, 677—693.

Lammens, E. H. R. R. (1985): A test of a model for planktivorous filter feeding by bream, *Abramis brama*. Environ. Biol. Fish. **13**, 286—296.

Leah, R. T., Moss, B., & Forrest, W. D. (1980): The role of predation in causing major changes in the limnology of a hypereutrophic lake. Int. Rev. ges. Hydribiol **65**, 223—247.

Linfield, R. S. J. (1984): The impact of zander (*Stizostedion lucioperca* (L.)) in the United Kingdom and the future management of affected fisheries in the Anglian region. EIFAC Tech. Pap. 42. Suppl. Vol. **2**: 353—362.

Patrick, B., & Haas, R. De (1971): Fin pulling as a technique for marking muskellunge fingerlings. Prog. Fish-Cult. **33**, 116—118.

Schlumpberger, W., & Schmidt, K. (1980): Vorläufiger Stand der Technologie zur Aufzucht von vorgestreckten Zandern (*Stizostedion lucioperca* L.). Z. Binnenfisch. DDR **27**, 284—286.

Shapiro, J., & Wright, D. I. (1984): Lakes restoration by manipulation: Round Lake, Minnesota, the first two years. Freshw. Biol. **14**, 371—383.

Steffens, W. (1981): Moderne Fischwirtschaft — Grundlagen und Praxis. Melsungen.

Svärdson, G., & Molin, G. (1973): The impact of climate on Scandinavian populations of the zander, *Stizostedion lucioperca* (L.). Rep. Inst. Freshw. Res. Drottningholm **53**, 112—139.

Swenson, W. A. (1977): Food consumption of walleye (*Stizostedion vitreum vitreum*) and sauger (*S. canadense*) in relation to food availability and physical conditions in Lake of the Woods, Minnesota, Shagawa Lake, and Western Lake Superior. J. Fish, Res. Bd. Can **34**, 1643—1654.

Tölg, J. (1981): Fortschritte in der Teichwirtschaft: Spezielle Methoden Hamburg, Berlin.

Willemsen, J. (1977): Population dynamics of percids in Lake IJssel and some smaller lakes in the Netherlands J. Fis. Res. Bd. Can **34**, 1710—1719.

Received December 9, 1986

Addresses of the authors:

W. L. T. Van Densen
Department of Fish Culture and Fisheries
P. O. Box 338
6700 AH Wageningen
The Netherlands.
M. P. Grimm
Organization for the Improvement of Inland Fisheries (OVB)
P.O. Box 433
3430 AK Nieuwegein
The Netherlands.

Limnologica (Berlin) 19 (1), 51—59, Juli 1988

Institut für Binnenfischerei, Berlin-Friedrichshagen, DDR

Fish Predation and Resource Reaction: Biomanipulation Background Data from Fisheries Research

By DETLEV BARTHELMES

With 7 Figures and 2 Tables

Key words: Impact of fish; Critical metabolic fish biomass concept; Pike-perch population dynamics; Induction of clear water

Abstract

From fisheries research it may be deduced that there are both of two situations in nature: Direct relationships between fish food species biomass and fish yield at one hand and inverse ones at the other. High amounts of fish food biomass are produced additionally with increasing fish feeding pressure starting from the low pressures of direct relationships and proceeding to the high pressures of inverse ones. This regulation capacity results in two reaches of the fish biomass that should be considered by water quality biomanipulation· The lower reach is characterized by the absence of marked fish-induced changes in the fish food biomass The upper reach is characterized by overgrazing of the food resource. It is tried to define dimensions of critical metabolic fish biomasses that are just sufficient to eradicate the prey populations discussed. There is evidence for natural fish populations of eutrophic lakes to generally overgraze large zooplankton. Pike-perch stocking of such lakes with the aim of reducing small cyprinids to below critical levels is not yet a safe measure because too much of the population ecology of both predator and prey species remained unknown or is not known in each special case. Cannibalism of the predator and inappropriate management of the stock can induce high instability that adds to the instabilities produced by *Aphanizomenon* on the primary production level and predatory invertebrates eating *Daphnia* on the secondary production level. Nevertheless, phases of increased transparency that sometimes seem to be stabilized by subsequent submerged vegetation growth can be produced this way (criterions of "Zandermüdigkeit"). This is also the case with extraordinary intense seine fishing Perhaps in future the combination of both measures will give better results. Research should proceed in this direction

1. Introduction

When in 1922 the International Association of Theoretical and Applied Limnology was founded by THIENEMANN and NAUMANN it was intended to give due consideration also to applied fields. With respect to fisheries this was well documented during the early congresses, but later on fisheries science was more or less underrepresented. So it happened that limnology and inland fisheries were not as fully connected with each other as desirable (RIGLER 1982). Today there is a real need to integrate the achievements of limnology into fisheries science at a higher rate. But the transfer of knowledge should also proceed from fisheries science to limnology at a sufficient and intensified rate. Lack of transfer from fisheries to limnology is especially apparent in the fast growing biomapulation sector where fish are of prime interest. It was in favour of the exchange and mutual integration that I wrote a book on the "Hydrobiological basis of freshwater fisheries" (BARTHELMES 1981), which included much of what is called now biomanipulation. Also the present paper tries to give some partly neglected data on how fishes influence their food resources which in turn influence other limnological parameters. Information of this kind is available since inland fisheries sience in Europe followed only one of two approaches. To model fish populations without considering the interconnections with other ecosystem compartments as practized in marine fisheries research for a long time was felt inappropriate under fresh water conditions, mostly because of the high number of isolated and small waterbodies (BARTHELMES 1984a). A more promising approach was to trace the special features of each waterbody from the fisheries level down to invertebrate fish-food production, primary production and nutrients. These thoughts governed the activities of freshwater fishery biologists from the beginning of the century onwards, and especially those of the so-called Friedrichshagen-school (SCHIEMENZ, WUNDSCH, SCHÄPERCLAUS). Later on, in the fifties, HRBÁČEK and his coworkers discovered the unexpected deep influence excerted by dense fish populations on structure and function of freshwater ecosystems. This became a leading idea and introduced the present-day views of biomanipulation (HRBÁČEK 1958, HRBÁČEK et al. 1961).

In summarizing, it may be pointed out that fisheries management in fact is mainly biomanipulation according to the definition given in the invitation to the first international conference on biomanipulation. Of course, the aims are different: Optimizing fish production versus optimizing water quality. But the internal functional relationships of the aquatic ecosystems profited by both management goals are identical.

2. The results of early fisheries research

The first results concerning the present topic emerged from carp-pond investigations as early as 1900 or so. It was then that WALTER questioned the validity of yield taxation ("Bonitierung") on the basis of bottom-dwelling fish food animal biomass (WALTER 1895, 1905). In his opinion there was no direct relationship between fish food biomass and fish yield. His view of an inverse relationship between the two variables is now beeing widely accepted (BARTHELMES 1986). In spite of such critical voices, yield prediction assuming direct proportionality between fish food biomass and fish yield was an important business during the first half of the century. There was a long-lasting controversy producing many investigations on the subject. In order to avoid a useless repetition on the biomanipulation level the rational underlying both views should be remembered: Below a certain threshold of fish feeding pressure and hence fish population density the losses of fish food biomass are compensated for by enhanced production of the organisms involved. With respect to bottom animals in carp ponds this was demonstrated by SCHÄPERCLAUS (1943). The same applies of course to zooplankton and its most important fraction for carp, the daphnids. If fish eat much more than approximately 25% of the *Daphnia* biomass found in carp ponds per day, the *Daphnia* populations will become extinct (SLOBODKIN 1959, HALL 1964, UHLMANN 1967, BARTHELMES 1969a, b). If the feeding pressure is lower, the *Daphnia* biomass will only be reduced to some equilibrium level. With further decreasing feeding rate the *Daphnia* biomass will soon arrive at a nearly untouched level. This is amplified by the Figs. 1—3.

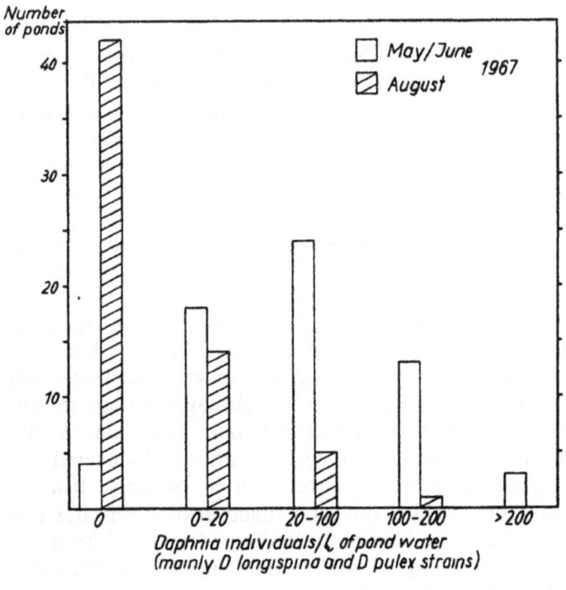

Fig 1. *Daphnia* abundance in 62 carp ponds stocked with one summer old carp (K_{0-1}) scattered over the whole territory of the GDR so as to give representative results. The carp fry stocked does not impact the *Daphnia* abundance shortly after stocking in May/June, but does so severely in August because of increased biomass and food demand. (From BARTHELMES 1969b).

It should be stressed, that in the case of one summer old carp stocks there was a negative relationship between August *Daphnia* abundance and yield, whereas in the case of the lowest stocking variant (stocking big one summer old carp at low densities in order to produce marketable fish in two years instead of three years) there was a positive one (BARTHELMES 1969b). Thus both relationships and transient situations may be encountered in nature and it is a matter of fish stock density, age structure, food preference, and also of cover availability (e.g. submerged vegetation) for prey organisms to hide, of prey productivity and other factors what equilibrium level of fish-food organisms is maintained in each special case.

Generally, in former decades the inverse relationship between fish-food organisms and fish biomass was much weaker than today in European waters. Enhanced eutrophication reduced the submerged vegetation greatly in lakes and ponds that hitherto provided shelter for fish food animals. Today the most important part of the bottom fish food animals living in warm epilimnic waters of lakes and ponds is much more available to fish, whereas with the former vegetation cover the availability was highly reduced, often to few percent (e.g. LUNDBECK 1927, HRUŠKA 1961, HANTGE 1962).

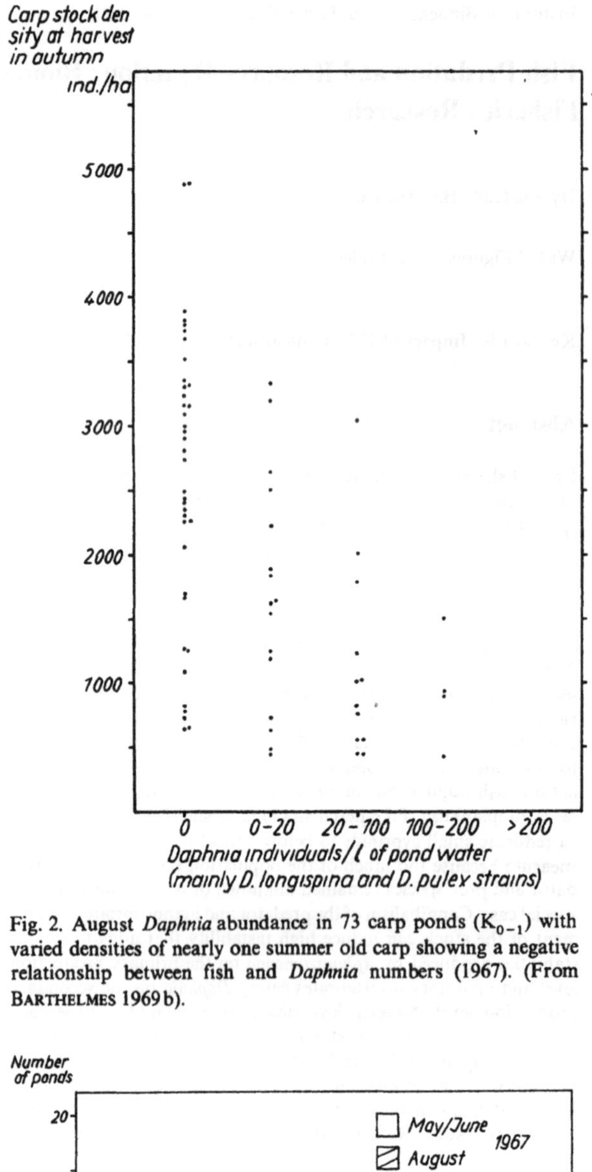

Fig. 2. August *Daphnia* abundance in 73 carp ponds (K_{0-1}) with varied densities of nearly one summer old carp showing a negative relationship between fish and *Daphnia* numbers (1967). (From BARTHELMES 1969b).

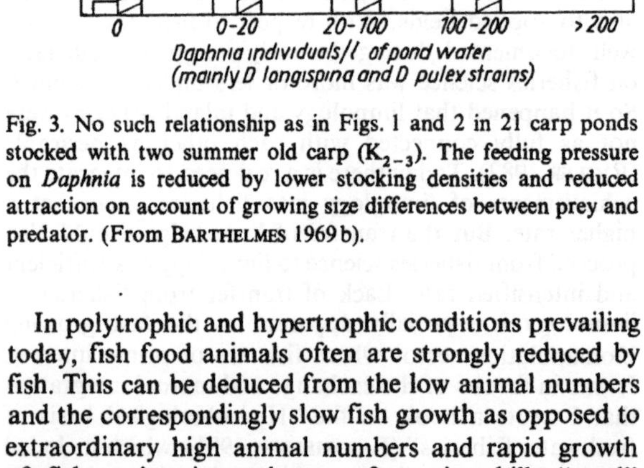

Fig. 3. No such relationship as in Figs. 1 and 2 in 21 carp ponds stocked with two summer old carp (K_{2-3}). The feeding pressure on *Daphnia* is reduced by lower stocking densities and reduced attraction on account of growing size differences between prey and predator. (From BARTHELMES 1969b).

In polytrophic and hypertrophic conditions prevailing today, fish food animals often are strongly reduced by fish. This can be deduced from the low animal numbers and the correspondingly slow fish growth as opposed to extraordinary high animal numbers and rapid growth of fish eating invertebrates after winterkill, "total" fishing or other catastrophes (e.g. SCHÄPERCLAUS 1947, PATALAS & ZAWISZA 1966, HRBÁČEK et al. 1961). On the other hand, stocking natural waters with invertebrate eating fishes often provides evidence for further reductions of fish food animals (see examples given in BARTHELMES 1981). This highlights once more the importance of production regulation by invertebrates mentioned above that never should be forgotten. An absolute figure of the magnitude of this regulation potential is available from carp-pond investigations: In fattening ponds that are characterized by Fig. 3 with respect to daphnids it amounts to approximately 2.55 t/ha · a, i.e. to nearly one half of the total animal food intake of the fish (BARTHELMES 1967).

3. An attempt to define critical fish biomasses

According to the preceding section several factors are decisive for the fish food animal level maintained. Fish biomass is only one of them and it is hard to determine exactly. Especially littoral fish fry densities are difficult to estimate. In spite of these problems it was tried to summarize the magnitudes hitherto known from G.D.R. fisheries research. The results are shown in Fig. 4. The critical fish biomass is intended to denote that magnitude which is just sufficient to eradicate the prey populations discussed. In order to enlarge the meaning of fish biomass it was converted to metabolic biomass using equation (1):

$$MW = W^{0.8} \cdot n \quad (1)$$

with MW = Metabolic biomass (kg); W ≙ Individual biomass (kg); n = Number of individuals in the weight-class considered.

This procedure corrects for size-dependent differences in metabolic rates and hence food demand and feeding pressure of 1 kg of fish biomass. In Fig. 4 the metabolic fish biomass is plotted against the productivity of the food organisms using the production/biomass ratio per year (P/B). As may be seen from the position of grass carp reducing macrophytes with P/B ratios slightly over 1 on one hand and silver carp reducing rotatoria with P/B ratios of about 50 on the other hand and the other cases lying somewhere in between there is an apparent direct relationship between the critical metabolic fish biomass and the productivity of the fish food organisms. This meets expectation. But the interference of the other factors mentioned is also apparent. The extinction of big *Daphnia* by different fish species under different conditions may serve as an example. The uppermost three quadrats refer to common carp in carp ponds (with additional grain feeding) and represent a summary of such data as shown in Figs. 1—3.

In the first year of carp rearing, fry (K_0) and fingerlings (K_v) are stocked at high rates (e.g. 30.000 K_v/ha). Therefore and because of the near optimum size relationship between the predator and its prey the critical metabolic fish biomass is lowest compared to the following age classes raised in carp-ponds. It amounts to 600—700 kg/ha. In the second year of carp rearing carp are stocked at a lower rate (e.g. 5.000 ind./ha) and the fish are already bigger. Thus the availability of *Daphnia* declines and the critical metabolic fish biomass ascends to 700—800 kg/ha. In the third year when the carp are reared to market size using stocking rates of 1000 ind./ha the process continues and the availability of *Daphnia* declines further. So the critical metabolic carp biomass raises to more than 1 t/ha. These relationships were found in carp-ponds without indigenous fish and their fry. Figures derived from lake investigations usually include the feeding pressure excerted by small fish and fry that are not accounted for. The magnitude of this error can be seen from the position of K_{1-2} carp in lakes (also additionally grain-fed) in relation to that of K_{1-2} carp in ponds (Fig. 4). Though the lake referred to was heavily fished by seine with the aim to reduce the indigenous fish stock drastically, its influence is still so strong as to lower the figure of the critical metabolic fish biomass from 700—800 kg/ha without trash fish in ponds to 400 kg/ha with such fish in lakes. According to expection the most important part of naturally occurring fish stocks with respect to *Daphnia* predation seem to be the fractions of small fish and fry that is hard to estimate in the field. This impairs the reliability of figures taken from field investigations and the utility of the critical fish biomass concept generally.

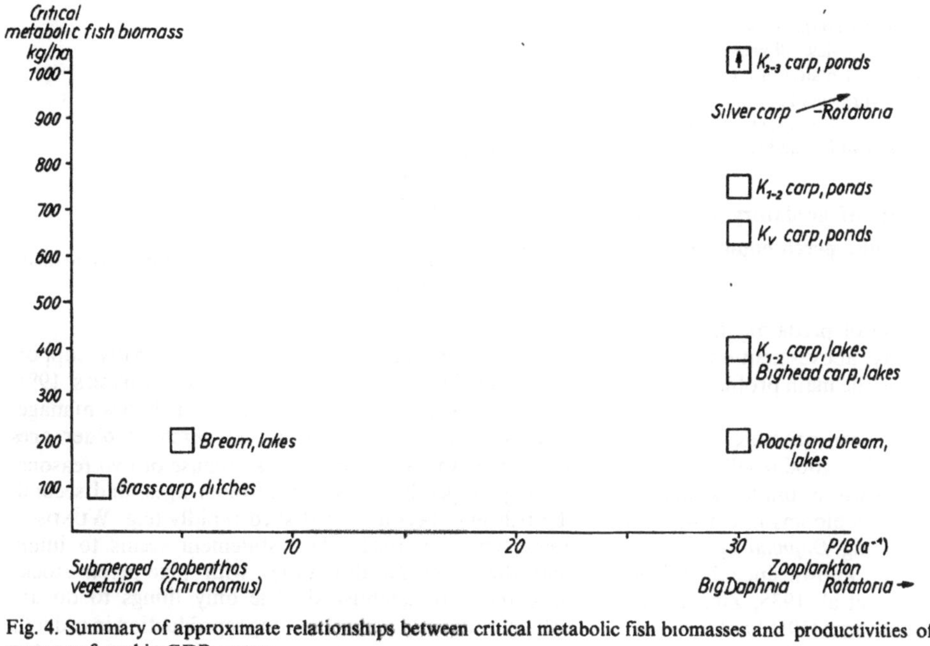

Fig. 4. Summary of approximate relationships between critical metabolic fish biomasses and productivities of food resources found in GDR-waters.

A further example of the importance of small fish and fry is provided by the position of bream reducing *Chironomus* in relation to a joint stock of roach and bream reducing *Daphnia*. According to the overall positive relationship between critical fish biomass and productivity of food resource one would suppose a higher position of the joint stock of roach and bream. But apparently roach and its offspring are very effective *Daphnia* predators. They reduce the joint critical biomass to the very level attained by bream alone when eating comparatively low productive *Chironomus*.

At present, with the common lack of information about the influence of small fish on the relationship discussed, it seems only possible to accept it as an error. This error renders all figures very rough ones that are gathered from lakes without special estimates of the smallest size-classes of fish. But, presently, a rough idea is perhaps better than no idea at all. The more so since limnological work on biomanipulation often could still profit from comparing the chosen experimental parameters with such rough magnitudes. For practical management purposes we use plots of permissible metabolic fish biomasses that resemble the critical metabolic fish biomass concept but proved successful so far in spite of the "fry error" inherent also here. Fig. 5 shows such a plot of permissible silver carp biomass as an example.

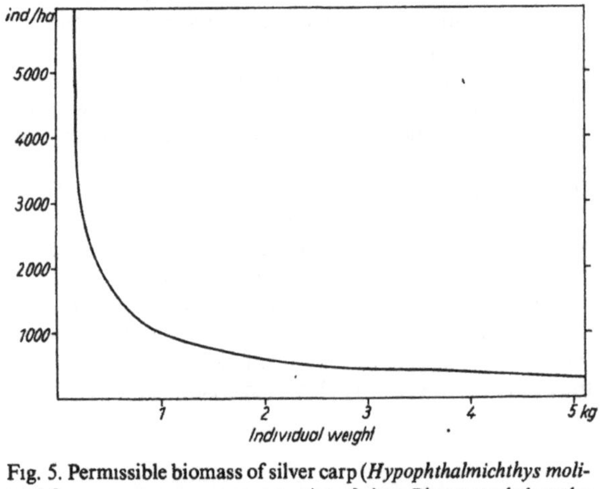

Fig. 5. Permissible biomass of silver carp (*Hypophthalmichthys molitrix*) from a water management point of view. Biomasses below the curve do not induce severe changes in plankton communities of highly eutrophic ponds. The curve represents 1t of metabolic silver carp biomass per ha. It seems to work also in lakes provided half this value is adopted of account of small fish additions.

4. Possibilities and restraints of predatory fish stock enhancement: New views of pike-perch population dynamics

The biomanipulation potential of predatory fish stocks is discussed at length in my above-mentioned book (BARTHELMES 1981). Repeating the main problems, it can be stated that

(a) dense *Daphnia* populations can promote blooms of non-edible algae and thereby offset good results of forage fish reduction. The best known example is that of *Daphnia*-induced *Aphanizomenon* blooms in carp-ponds. The combined occurrence of big *Daphnia* species and *Aphanizomenon* in carp-ponds is known since the thirties (e.g. WEIMANN 1938, WUNDER et al. 1935, ZIEGELMEIER 1940) and was referred to by WURTZ (1958) as "famous". SHAPIRO and his co-workers, proceeding parallel and without knowledge of these results, arrived also at this problem (SHAPIRO et al. 1975, SHAPIRO et al. 1982).

(b) *Daphnia* populations can be reduced by predators other than fish, e.g. by *Leptodora*, *Chaoborus*, Cyclopidae (BARTHELMES 1981) or *Mysis* (LASENBY & FÜRST 1981). In such cases the reduction of plankton-eating fish might not work.

(c) at the fish population level there can evolve some regulation in response to predation. For instance pike-perch prefers small forage fish. So it can reduce tall fish substantially (e.g. roach) while advancing compressed, deep-bodied ones (e.g. bream) (TESCH 1965, BATHELMES 1981, BARTHELMES & PAETSCH 1985), that eat *Daphnia* too, at least during the first three years of their lives.

Therefore, no clear-cut relationship between predatory fish stock density and water transparency is to be expected.

Often pike-perch (*Stizostedion lucioperca*) is supposed to work best of all rapacious fish species encountered in European eutrophic waters. Pike-perch prefers highly eutrophic conditions as opposed to pike (*Esox lucius*) and perch (*Perca fluviatilis*) that additionally exhibit strong self-regulation by cannibalism. Furthermore, pike-perch can reduce the density of smaller forage fish efficiently. Figs. 6 and 7 demonstrate this for 42 Polish pike-perch lakes according to BONAR (1977).

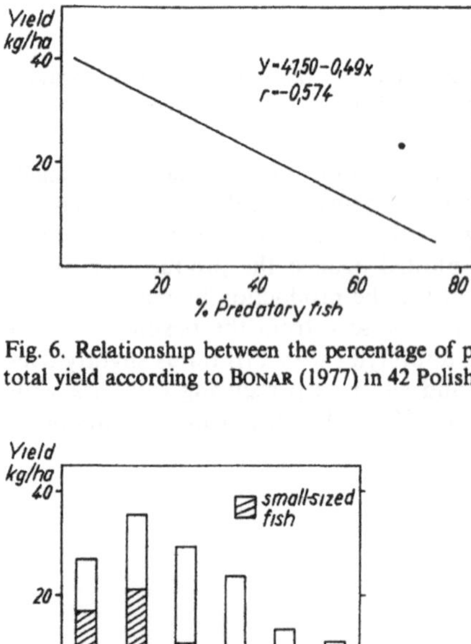

Fig. 6. Relationship between the percentage of predatory fish and total yield according to BONAR (1977) in 42 Polish pike-perch lakes.

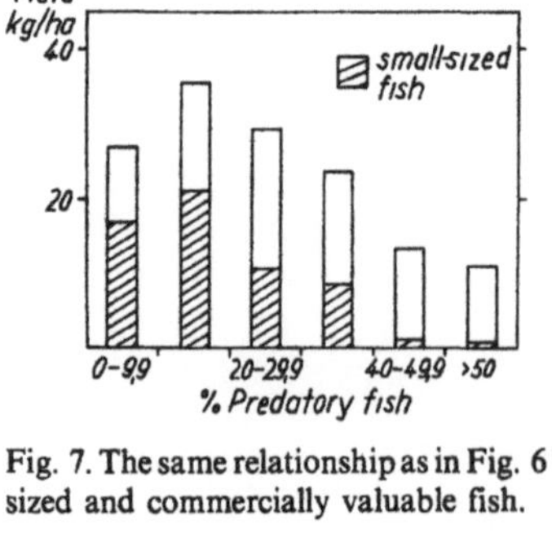

Fig. 7. The same relationship as in Fig. 6 but yield divided into small-sized and commercially valuable fish.

Also other observations made occasionally support this view (POPOVA & SYTINA 1977, BARTHELMES 1981, LINFIELD 1982). Textbooks on inland fisheries management recommend to protect pike-perch to older ages than other valuable fish species because of two reasons: First, pike-perch grows well at older ages, and second, the fish may become overfished rapidly (e.g. WUNDSCH 1960, STEFFENS 1981). This statement seems to imply that there are no difficulties with pike-perch stocks once they are established. The only things to do are fishing cautiously and promoting reproduction in order to get rich recruitment. Recruitment is admitted to fail sometimes when food conditions for fry are poor. But this may be overcome by stocking advanced fry.

This simple view that certainly encourages pike-perch stocking experiments for biomanipulation purposes is not in agreement with some recent results. In fact, the situation is much more complicated and this was evidenced already by former observations. BAUCH (1952) mentions for instance the phenomenon of "Zandermüdigkeit", i.e. a strong decrease of pike-perch yields accompanied by clearer water and recovery of submerged vegetation, of pike, tench and bream in pike-perch lakes (see also SCHÄPERCLAUS 1960). Such events were interpreted in terms of weather conditions and hydrological changes. Therefore they were accepted as natural events that could not be influenced. According to new results this view is quite misleading. Instead, the causes can be derived from the following puzzle: Pike-perch stocks tend to be underfished with the above-mentioned recommendation in mind. This favours prey fish overexploitation by the pike-perch stock — the very aim of biomanipulation — and in accordance clearer water and enhanced growth of remaining fish that profit from dwindling feeding pressure against bottom and plankton fish food animals (BARTHELMES 1986). At the same time the improved underwater light climate allows submerged vegetation and the accompanying fauna including pike and tench to recover. That all are the symptoms of "Zandermüdigkeit". But now a new factor comes into operation that was strongly neglected by early fisheries biology: Cannibalism of pike-perch in case of food lack. This factor was found to operate in all populations investigated thouroughly and is now being reported from different localities (e.g. VIRBICKAS et al. 1974, BIRO 1975, POPOVA & SYTINA 1977, NAGIEC 1977, COLETTE et al. 1977, WINKLER 1980, DENSEN & GRIMM 1988). It applies to the North American *Stizostedion vitreum*, too (FORNEY 1980). There is no doubt about the general importance of cannibalism in pike-perch populations now, which also is a new experience compared to textbook teaching. Certainly, it was also involved in spectacular pike-perch declines in some G.D.R. lakes, as for instance in Lake Schweriner Innensee (decline from a commercial yield of 19,4 kg/ha in 1975 to 1.7 kg/ha in 1983, WICHMANN 1985), Lake Müritz (BARTHELMES & PAETSCH 1985) and also in Lake Großer Müggelsee, which is one of 10—15 experimental lakes that we investigate simultaneously for limnological and fisheries parameters. Lake Großer Müggelsee yielded about 12 kg/ha pike-perch to commerical fishermen the years before 1978 (BARTHELMES & WALDOW 1978). Then, during a few years, the yield dropped to below 1 kg/ha. Simultaneously, the smelt (*Osmerus eperlanus*) representing a main prey of pike-perch became nearly extinct, while the growth rate of roach and bream greatly improved (BARTHELMES 1983a). This year (1986), the smelt reappeared in moderate numbers as shown by smelt trawl catches, whereas commercial pike-perch catches and thus pike-perch stock continued to be low. During the time of declining stock size the pike-perch was extremely vulnerable to angling. Similar coupling of rapid decline of pike-perch yield with angling success is reported by practical fishermen from other waters including the above-mentioned lakes Müritz and Schweriner Innensee. These observations are in agreement with published data from Oneida Lake and Tjeukemeer (FORNEY 1980, DENSEN & VIJVERBERG 1982). They confirm the mechanism inherent in this decline once more independently.

Once pike-perch stocks have done their work, i.e. eaten most of the small fish present, they will decline and given way to a new pulse of small fish. Thus, pike-perch populations often will oscillate between the extremes of the BONAR-relationship which no longer should be taken for static. The time elapsing between pike-perch peaks may be variable. Anyway, it is in the magnitude of years. But according to the old observations on "Zandermüdigkeit" that are felt to mostly involve the presented pattern of events the duration of the cannibalistic pike-perch depression is sometimes only short, i.e. 2 or 3 years (BAUCH 1952). Other cases as for instance of Lake Müritz and of Lake Großer Müggelsee exhibit very prolonged depressions that imply no good predictability of what a reduced pike-perch population will produce in future.

One important observation made at several lakes is nearly total failure of rich one summer old age-classes of pike-perch to recruit to harvestable fish. In our own experimental lakes this was well documented by Lake Wriezensee (for main features of this lake see BARTHELMES 1983b). The same was reported for Lake Grimnitzsee (SCHMIDT, pers. comm.). In former years it was a common experience that stocking one summer old pike-perch in lakes that seemed to be suited for this fish gave only poor results with respect to recruitment. RAHN (1958) reported a mean recruitment rate of only 13% in G.D.R. lakes. This figure does not yet give the full truth since total catch of pike-perch was correlated with total stocking. Thus, natural reproduction is included and stocking yielded still less.

According to an old practical experience stocking with pike-perch is only promising if the otherwise suited lakes are not yet populated at all, but not necessarily if existing populations are wanted to produce (and eat) more. An anlysis of 125 stocking experiments with the North American *Stizostedion vitreum* made by LAARMAN (1978) strongly confirms this view. Only 5% of the stocking experiments were successful in lakes with native self-reproducing stocks as opposed to 48% in lakes without the species that is very near related to our *Stizostedion lucioperca*. This difficult situation may perhaps explain the widely varying opinions about the individual size the stocking material should have: In our modern textbook on fisheries management the above-mentioned failure of one summer old pike-perch to recruit led to the supposition that advanced fry might succed (STEFFENS 1981, see also DENSEN & GRIMM 1988). In contrast, DAHL (1982) in summarizing 128 case-histories, concluded that stocking only should be done with fish of age two or three and of 20—30 cm length. But according to LAARMAN (1978) number and size of the fish to be stocked are less important. Instead, the prevailing ecological conditions in the water to be stocked are most decisive. Just this can be deduced also from KIRILLENKO (1984). Perhaps this conclusion is nearer to the truth than all others because it correlates well with the pattern described above.

Taken all the observations together the pike-perch populations emerge as sensitive units that are hard to manage satisfactorily with the present knowledge and instruments of fisheries management (see also SVÄRDSON & MOLIN 1973). The prey populations can regulate the predator populations (FORNEY 1980). Accordingly, the introduction of higher minimum size limits produced different results (e.g. SCHLUMPBERGER 1977, WINKLER 1980). Such limits might be useful only with frequent corrections for population parameter changes (SERNS 1978). The practical performance of such a tool remains still for elaboration. Presently, methodical difficulties regarding the in situ estimation and tracking of size and structure of both the predator and prey populations prevent deeper knowledge. This "black box" situation also refers to practical stocking measures when the ecological situation is not properly known. At present one very important precondition for pike-perch stockings or populations to flourish is thought to be a plentiful supply of suitable-sized prey. Morphological pecularities force pike-perch to eat narrow prey sizes that are defined by its own size. Because of these narrow size limits of useful prey the prey size spectrum, which is necessary to avoid gaps that are followed by starvation and cannibalism of the predator, ought to be extraordinarily rich. Thus, to ensure good food conditions of age 0 pike-perch throughout the whole year, forage fish species spawning late in summer or spawning in portions are necessary. With respect to older ages small fish species or stunted populations of fish species that grow slowly to bigger sizes are important (roach for instance). Biomanipulation necessitates a balance between the reduction of prey populations to yield below critical biomasses of all zooplankton-eating fish on one hand and the maintenance of sufficient food for the predator at the other. From present knowledge of fisheries management it is not yet clear whether a balance of this sort can be achieved that satisfies the biomanipulation goal.

From the restricted present-day knowledge it may be added, that sometimes such instability as exemplified by "Zandermüdigkeit" does not develop. This is the case with lakes that are fairly intense fished for pike-perch or where the free niche produced by pike-perch predation is occupied at the same or a delayed rate by bream. Bream is only vulnerable to pike-perch predation during a comparable short time span on account of its extraordinary body depth. Such a situation presumably was found to exist in one of our experimental lakes, Lake Bauernsee. The population densities were estimated from mark and recapture experiments and subsequent rigorous seine fishing. The results are summarized in Table 1 together with those from a near by reference lake (Lake Eiserbuder See). In Lake Bauernsee, the pike perch population is very large as compared to the reference lake and other waters (Balaton 10—20 kg/ha after BIRO 1977, a mean of 6 kg/ha in Finland after LIND 1977, 17—21 kg/ha in Lake Jesiorak after NAGIEC 1978 etc.). This is also the case with the bream population (for comparison see total fish biomasses of lakes listed in BARTHELMES 1981). The fish populations of Lake Bauernsee persisted in the very composition shown in Table 1 at least 5 years, because during that time no substantial catches of native fish were made on the lake. Apparently, the pike-perch ate predominantly small bream that were supplied continuously by the adult stock. Such situations of adult stock protection from predation may lead to stability (MURDOCH & OATEN 1975). But this stability, though reached at a very high population level, did not work in a biomanipulation sense. This is shown by Table 2 that exhibits nearly identical limnological parameters in both of the lakes. The differences in abundance of cladocera and copepoda are also interpreted as relatively small with respect to the rough methods chosen.

Lake Bauernsee and Lake Eiserbuder See do not differ in these parameters but they do so in fishing pressure, Lake Eiserbuder See being the most intensely and continuously fished one. The reduced total fish biomass and especially the reduced bream biomass in Lake Eiserbuder See which otherwise compares well with Lake Bauernsee only induced a much higher benthos biomass and hence a much better growth of bream and roach (BARTHELMES 1986).

Concluding, it should be stressed again that knowledge is not yet sufficient to answer all questions. So it is not clear, for instance, what situation would have developed in Lake Bauernsee with different size distributions of pike-perch and different fishing intensities for bream (see following section).

Table 1. Fish biomasses in lakes Bauernsee and Eiserbuder See in 1983

Species and length interval		Bauernsee	Eiserbuder See
		(kg/ha)	
Bream	(> 50 cm)	0	20
Bream	(35—50 cm)	0	present
Bream	(25—35 cm)	600—1200	58
Roach	(20—24 cm)	0	16
Roach	(15—20 cm)	50	54
Carp	(old fish)	30	50
Perch	(21—30 cm)	present	4
Pike-perch	(> 40 cm)	60—120	9
Pike	(30—50 cm)	5	12
Eel		not included	

Table 2. Mean limnological parameters of lakes Bauernsee and Eiserbuder See in 1983

		Bauernsee, 19 ha, z_{max} 3.6 m, not stratified		Eiserbuder See, 37 ha, z_{max} 6 m, partly stratified
Total phosphorus	(mg/m³)	335		287
Chlorophyll a	(mg/m³)	87.3		90.6
Phytoplanktón	(cells/ml)	321 000		368 000
Seston	(mg/l)	30 5		27 6
Secchi depth	(m)	0.47		0.53
Cladocera	(ind./l)	737	similar size spectra	238
Copepoda	(ind./l)	396		733
Rotatoria	(ind./l)	4623		2525
Benthos biomass	(g/m²)	1		13
Total fish biomass (fish ≧ 15 cm)	(kg/ha)	745—1405		223

5. Observations about increased transparency of lakes induced by fisheries management other than predator enhancement

If benthos biomass is excluded from further considerations because of its peculiar availability features and interest centers only on zooplankton then lake fish biomass is felt to be generally above the critical metabolic level discussed in chapter 3.

The already mentioned change from small crustaceans to bigger ones following winterkill and other catastrophes in small lakes is in line with this view. The small species-dominated (or lucid species-dominated) zooplankton communities of eutrophic lakes did not change their character since old ages except for evolutionary alterations (LIEDER 1983). With the background information given this can be interpreted as evidence for strong fish feeding pressure existing already before man's fishing and management activities appeared. Such an interpretation that means naturally occurring metabolic fish biomasses above the critical level with respect to large *Daphnia* species as defined above is in agreement with the carrying capacity concept of CARLANDER (1966). It implies that fish populations of natural waters generally use the food resources in full. This may well involve some succession from larger to smaller zooplankton according to the availability sequence discussed above for common carp.

The impact of fish biomass reduction by fishing apparently was small in former days. Of course, if fishing is done continuously and at a small rate so as to take the surplus production month by month no severe disturbance of the natural fish biomass level is to be expected. During the last decades an other technology has developed in many countries. It is characterized by more intermittent but strengthened fishing activities. In the extreme, such small lakes are only fished once every few years but then total cropping is wanted. Limited gear efficiencies confine this technology to smaller lakes and also at such lakes it is not always successful. But there is much effort to reach this aim in connection with stocking common carp, silver carp and bighead carp. The main gear type used for total fishing is the seine, expecially large carp seines or specialized purse seine-like nets with electroshocking devices mounted on the underline. Apparently, with such gear and especially with such gear operated repeatedly it happens sometimes to get the total fish biomass reduced to below critical levels. This is what we observed in several of our experimental lakes. Thus, Lake Eiserbuder See developed blooms of big *Daphnia* that increased transparency following a new reduction of the overall fish biomass by seining in 1984. The biomass of fish $\geqq$ 15 cm long was as low as 100 kg/ha or so. The share of bream was reduced still further in comparison to the 1983 values shown in Table 1, because bream is much better retained by the net than the torpedo-shaped roach. In this way, severe fishing may withdraw or invert the pike-perch induced fish size community structure, speaking in favour of a possible complementary use of both principles in future. As mentioned, the example of Lake Eiserbuder See is introductory to a series of observations that show intense seine fishing to produce "clear water" otherwise highly eutrophic waters. But this "clear water" again is transitory at present, with no additional measures taken. It occurs mainly in May and June after "total fishing" the preceding autumn. In midsummer, it is clouded by blue-greens, especially *Aphanizomenon flos-aquae*, if the big *Daphnia* persist. The situation then resembles very much that of K_{2-3} carp ponds. It is most wide-spread in shallow, highly eutrophic lakes that are managed similar to carp ponds in that the indigenous fish stocks are reduced to very low figures by extreme fishing pressure including year by year repetition. Then K_2 carp are stocked at medium densities. In accordance with Fig. 3 the densities applied do not reduce the *Daphnia* stocks occurring because of the native fish stock decimation. An example of this kind is given in my book on the basis of Lake Seelübber See from our experimental lakes (BARTHELMES 1981, p. 180/181). Proper management and especially the extreme seine fishing pressure are essential for this situation to occur. Lakes that are not well suited for seine fishing also are not well suited for this kind of management. A pronounced shortcoming may originate in such lakes from unproportionate reduction of the big predatory fish, especially pike, which are caught at higher rates also with an inefficiently operated gear in comparison to non-predatory fish. Thus sometimes in such lakes the opposite of what originally was wanted takes place: An increase of non-predatory cyprinid fish that compete with carp and make good management results impossible. An example of this kind was provided by Lake Galenbecker See, which I discussed at length in replying to a voice claiming carp stocking of lakes to result necessarily in further eutrophication (BARTHELMES 1984b, KALBE 1984). Apparently, the most essential precondition of carp stocking in lakes is often neglected by non-professionals in favour of stressing the fact of stocking fish. This attitude produces a picture far from what carp management of lakes is in reality: A very strong reduction of native fish stocks and a subsequent stocking of much smaller numbers and biomasses of carp in order to realize rapid individual growth giving an overall net reduction of fish biomass and *Daphnia* eating age classes.

At last it shall be mentioned that exceptional prolonged clear water phases seem to be also possible as a consequence of intense fishing. This was observed with bighead carp as the species stocked. The same intense fishing given as with common carp the induced clear water under such conditions may become stabilized by subsequent growth of submerged vegetation. Perhaps in future it will be possible to artificially create this way clear and vegetation rich tench lakes starting from turbid shallow lakes void of soft vegetation. But at present it is not yet sure whether the reproduction of such situations will be possible. Much work is felt still necessary to understand the processes involved in order to use them in practical management. This seems also true of most biomanipulation procedures in general.

6. Summary

According to the definition given in the invitation to the first international conference on biomanipulation, fisheries management is mainly biomanipulation apart from the different management goals. From early and modern fisheries research it may be deduced that there are both of two situations in nature: Direct relationships between fish food organism biomass and fish yield at one hand and inverse ones at the other. Inverse relationships apply mainly to large zooplankton, especially *Daphnia* species, direct ones mainly to zoobenthos biomass in the presence of submerged vegetation reducing the availability of the zoobenthos to fish. Increased eutrophication induces a shift in favour of inverse relationships High amounts of fish food biomass are additionally produced with increasing fish feeding pressure starting from the low pressures of direct relationships and proceeding to the high pressures of inverse ones This regulation capacity produces two reaches of fish biomass that should be considered by water quality biomanipulation: The lower reach is characterized by absence of a visible negative impact on the fish food biomasses and hence stocking is neutral within this reach with respect to water quality. The upper reach is characterized by overgrazing of the food resource. It is tried to define the borderline by the aid of magnitudes of critical metabolic fish biomasses. There is evidence for natural fish populations of eutrophic lakes to generally overgraze large zooplankton. Pike-perch stocking of eutrophic lakes with the aim of reducing small cyprinids to below critical levels is not yet a safe measure because too much of the population ecology of both predator and prey species remained unknown or is not known in the special case. Cannibalism of the predator and inappropriate management of the stock emerging from lack of knowlege can induce high instability that adds to the instabilities produced by *Aphanizomenon* on the primary production level and predatory invertebrates eating *Daphnia* on the secondary production level. Nevertheless, transient phases of increased transparency that sometimes show some stabilization by subsequent submerged vegetation growth can sometimes be produced this way (criterions of „Zandermüdigkeit"). This is also the case with extraordinary intense seine fishing. Perhaps in future the combination of both measures will give better results. Research should proceed in this direction.

Zusammenfassung

Aus Erkenntnissen der Regulation von Fraßdruck und Beutedichte bei Fischen werden kritische Hinweise zur Biomanipulation mit Hilfe des Einsatzes von Zandern gegeben (Kannibalismus, Auftreten von *Aphanizomenon*, Zunahme planktivorer Invertebraten). Fischereiliche Gesichtspunkte der Biomanipulation werden diskutiert.

References

Barthelmes, D. (1967): Schätzung der Naturnahrungsverhältnisse im Karpfenteich als Grundlage zur Verbesserung der Kondition der Karpfen und zur Steigerung der Erträge. Dt Fischereiztg. **14**, 330—343

— (1969a): Möglichkeiten einer industriemäßigen Naturnahrungsproduktion in Kombination mit der industriemäßigen Karpfenproduktion. Z. Fischerei **17**, 117—125.

— (1969b): Beziehungen zwischen Bewirtschaftung sowie Verbreitung und Massenwechsel der Gattung *Daphnia* in den Karpfenteichen der Deutschen Demokratischen Republik. Z. Fischerei **17**, 177—203

— (1981). Hydrobiologische Grundlagen der Binnenfischerei. Jena, Stuttgart, 252 p.

— (1983a). Das Wachstum der Fische im Großen Müggelsee und die fischereiliche Bewirtschaftung des Sees. Vortrag Winterschulung PGB Berlin.

— (1983b): Effects of eutrophication and fisheries management on fish faunas of cyprinid lakes. Roczniki Nauk Roln Ser. H **100** (3), 31—44.

— (1984a): Fischereiwirtschaft. In. Schubert, R., Lehrbuch der Ökologie. Jena, 529—535.

— (1984b): Zur Frage der Karpfenintensivbewirtschaftung von Seen und Eutrophierung. Acta hydrochim. et hydrobiol. **12**, 153—161.

— (1986): Die fischereiliche Bonitierung von Seen aus heutiger Sicht. Vortrag VI. Wiss. Nutzfischkonferenz, Güstrow.

— & Paetsch, B. (1985): Zur langjährigen Entwicklung der Fischbestände in der Müritz sowie zur Bewirtschaftung von Gewässern mit ähnlicher Bestandsstruktur. Z. Binnenfischerei DDR **32**, 171—174.

— & Waldow, H. (1978): Langfristige biologische Veränderungen im Großen Müggelsee (Berlin), einige Kennzahlen des jetzigen Zustandes und bisherige fischereiliche Auswirkungen. V. Zusammenfassende fischereiliche Wertung der beobachteten biologischen Veränderungen im Großen Müggelsee. Z. Binnenfischerei DDR **25**, 209—214.

Bauch, G. (1952): Welche Gewässer eignen sich für die Bewirtschaftung mit dem Zander (*Lucioperca sandra*)? Fischen u. Angeln **4**, 100—105.

Biro, P (1975): Observations on the fish production of Lake Balaton Symp. Biol. Hung. **15**, 273—279

— (1977): Effects of exploitation, introductions and eutrophication on percids in Lake Balaton. J. Fish. Res. Bd. Can. **34**, 1678 to 1683

Bonar, A. (1977): Relations between exploitation, yield and community structure in Polish pikeperch (*Stizostedion lucioperca*) lakes, 1966—71. J. Fish. Res Bd. Can. **34**, 1576—1580.

Carlander, K. D. (1966): Relationship of limnological features to growth of fishes in lakes. Verh. Int. Verein. Limnol. **16**, 1172—1175

Colette, D. B., Ali, M. A., Hokamon, K. E. F., Nagiec, M., Smirnov, S. A., Thorpe, J. E., Weatherley, A. H., & Willemsen, J. (1977): Biology of the percids. J. Fish. Res. Bd. Can. **34**, 1890—1899.

Dahl, J. (1982): A century of pikeperch in Denmark. Symp. 12. EIFAC Sess., Budapest, 10p.

Densen, W. L. T. van & Grimm, M. P. (1988): Possibilities for stock enhancement of pikeperch (*Stizostedion lucioperca*) in order to increase predation on planktivores. *Limnologica* **19** (1), 45—49.

— & Vijverberg, J. (1982): The relations between O+ fish density zooplankton size and the vulnerability of pikerperch, *Stizostedion lucioperca*, to angling in the Frisian lakes Hydrobiologia (The Hague) **95**, 321—336.

Forney, J. L. (1980): Evolution of a management strategy for the walleye in Oneida Lake, New York. N.Y. Fish and Game J. **27**, 105—141.

Hall, D. J. (1964): The dynamics of a natural population of *Daphnia*. Verh. Int. Verein. Limnol. **15**, 660—664.

Hantge, E. (1962): Untersuchungen an Chironmiden aus Karpfenteichen. Arch. Hydrobiol. **58**, 309—338.

Hrbáček, J. (1958): Typologie und Produktivität der teichartigen Gewässer. Verh. Int. Verein Limnol. **13**, 394—399

— Dvořakova, M., Korinek, V, & Prochazkova, L (1961): Demonstration of the effect of the fish stock on the species composition of zooplankton and the intensity of metabolism of the whole plankton association Verh. Int. Verein. Limnol. **14**, 192—195.

Hruška, V (1961): An attempt at a direct investigation of the influence of the carp stock on the bottom fauna of two ponds. Verh. Int. Verein. Limnol. **14**, 732—736.

Kalbe, L. (1984): Karpfenintensivbewirtschaftung von Seen und Eutrophierung. Acta hydrochim. et hydrobiol. **12**, 145—152.

Kirillenko, L. W. (1984): Besetzen von Seen mit Zandern. Rybnoje chozj. **6**, 32 (russ.).

Laarman, P. W. (1978): Case histories of stocking walleyes in inland lakes, impoundments, and the Great Lakes — 100 years with walleyes. Amer. Fish. Soc. Spec. Publ. **11**, 254—260.

Lasenby, D. C., & Fürst, M. (1981): Feeding of *Mysis relicta* Loven on macrozooplankton. Rep. Inst. Freshw. Res. Drottningholm **59**, 75—80.

Lieder, U. (1983): Die Arten der Untergattung *Eubosmina* Seligo, 1900 (Crustacea: Cladocera, Bosminidae). Mitt. Zool. Mus. Berlin **59**, 195—292

Lind, E. A. (1977): A review of pikeperch (*Stizostedion lucioperca*), Eurasian perch (*Perca fluviatilis*) and ruff (*Gymnocephalus cernua*) in Finland. J. Fish. Res Bd. Can. **34**, 1684—1695.

Linfield, R S. J (1982): The impact of zander *Stizostedion lucioperca* (L.) in the United Kingdom and the future management of affected fisheries in the anglian region. Symp. 12. EIFAC Sess., Budapest, 16 p.

Lundbeck, J. (1927): Der „FB-Koeffizient" für Teiche Z. Fischerei **25**, 553—575.

Murdoch, W. W., & Oaten, A. (1975): Predation and population stability. Adv. Ecol Res. **9**, 1—125.

Nagieg, M. (1977): Pikeperch (*Stizostedion luciperca*) in its natural habitats in Poland. J. Fish. Res. Bd. Can. **34**, 1581—1585.

— (1978): Dynamik einer intensiv befischten Zanderpopulation (*Stizostedion lucioperca*) im Jeziorak-See in Masuren. Zeszyty nauk. Akad. rolniczotechn., Ochrona Wod i Rybactwo srodladowe **8**, 119—176 (poln.).

Patalas, K., & Zawisza, J. (1966): Limnological research in the fishery management of lakes in Poland. Verh. Int. Verein. Limnol. **16**, 1161—1171.

Popova, O. A., & Sytina, L. A. (1977): Food and feeding relations of Eurasian perch (*Perca fluviatilis*) and pikeperch (*Stizostedion lucioperca*) in various waters of the U.S.S.R. J. Fish. Res. Bd. Can. **34**, 1559—1570.

Rahn, J. (1958) Über Möglichkeiten und Grenzen der Steigerung des Feinfischertrages unserer Seen- und Flußfischerei durch Besatzmaßnahmen. Dt. Fischereiztg. **5**, 42—49.

RIGLER, F. H. (1982): The relation between fisheries management and limnology. Trans. Amer. Fish. Soc. **111**, 121—132.

SCHÄPERCLAUS, W. (1943): Der Einfluß verschiedener Faktoren auf die Mengenentfaltung der Chironomidenlarven am Teichboden. Arch. Hydrobiol. **40**, 493—524.

— (1947): Lebensbedingungen und Lebensfähigkeit der Fische in einem eutrophen Flachlandsee bei Eisbedeckung im Winter. Forsch. u. Fortschritte **21/23**, 182—185.

— (1960): Fischereiwirtschaft in Seen und Flüssen. In: WUNDSCH, H. H., Fischereikunde. 2. Aufl. Radebeul, Berlin, 154—194.

SCHLUMPBERGER, W. (1977): Analyse und Ergebnisse einer Mindestmaßänderung für den Zander (*Stizostedion lucioperca* (L.)) in Küstengewässern der mittleren Ostsee. Z. Binnenfischerei DDR **24**, 7—13.

SERNS, S. L. (1978): Effects of a minimum size limit on the walleye population of a northern Wisconsin lake. Amer Fish Soc Spec. Publ. **11**, 390—397.

SHAPIRO, J., LAMARRA, V., & LYNCH, M (1975): Biomanipulation. An ecosystem approach to lake restoration. Contribution **143**, Limnol. Res. Centre, Univ. Minnesota, 1—32.

— FORSBERG, B., LAMARRA, V , LINDMARK, G., LYNCH, M , SMELTZER, E., & ZOTO, G. (1982): Experiments and experiences in biomanipulation Interim. rep. **19**, Limnol. Res. Centre, Univ. Minnesota, Minneapolis, 251 p.

SLOBODKIN, L. B. (1959): Energetics in *Daphnia pulex* populations. Ecology **40**, 232—243.

STEFFENS, W. (1981): Industriemäßige Fischproduktion. 2 Aufl. Berlin, 376 p.

SVÄRDSON, G , & MOLIN, G. (1973): The impact of climate on Scandinavian populations of the sander, *Stizostedion lucioperca* (L) Rep Inst. Freshw. Res. Drottningholm **53**, 112—139.

TESCH, F. W. (1965): Wie Zander und Blei den Barschbestand verdrängen Allg. Fischerei-Ztg. **90**, 726—728.

UHLMANN, D (1967): Beitrag zur Limnologie saprotropher Flachgewässer. Arch. Hydrobiol. **63**, 1—85.

VIRBICKAS, J., GERULAITIS, A., MISIUNIENE, D., & SINEVICIENE, D. (1974): Biologie und Fischerei des Zanders in den Gewässern Litauens. Inst. Zool. a. Parasitol. Akad Sci. Lit. SSR, Vilnius, 279 p. (russ.).

WALTER, E. (1895). Über die Möglichkeit einer biologischen Bonitierung von Teichen. Allg. Fischerei-Ztg. **10**, 95—99, 120—125, 139—142.

— (1905): Über die Fragen der Bonitierung und der Nahrungsuntersuchung. Dt. Fischereiztg. **8**, 160—161, 176—178, 188 bis 192, 201—204.

WEIMANN, R. (1938): Planktonuntersuchungen im niederschlesischen Karpfenzuchtgebiet Z. Fischerei **36**, 109—184.

WICHMANN, F. (1985): Vorschläge für eine effektive Bewirtschaftung der Massenfischbestände des Schweriner Sees mit dem Ziel ihrer Nutzung für die menschliche Ernährung. Humboldt-Univ. Berlin, Sekt. TPV, Fachr. Fischprod., Diplomarbeit, 61 p.

WINKLER, H. (1980): Untersuchungen zur Fischerei und Biologie des Zanders (*Stizostedion lucioperca* (L.)) in einem hocheutrophen brackigen Küstengewasser der westlichen Ostsee. Diss., Wilhelm-Pieck-Univ Rostock, 145 p.

WUNDER, W., UTERMÖHL, H., & OHLE, W. (1935): Untersuchungen über die Wirkung von Superphosphat bei der Düngung großer Karpfenteiche. Z. Fischerei **33**, 555—613.

WUNDSCH, H. H. (1960). Fischereikunde. Radebeul, Berlin, 351 p.

WURTZ, A. (1958). Peut-on concevoir la typification des étangs sur les memes bases que celles des lacs? Verh. Int Verein. Limnol. **13**, 381—393.

ZIEGELMEIER, E. (1940): Die qualitative und quantitative Verteilung des Zooplanktons in einigen großen Fischteichen der Bartschniederung mit besonderer Berücksichtigung der Cladoceren und Copepoden. Arch Hydrobiol. **36**, 495—551.

Received January 14, 1987

Address of the author:

Dr. habil. DETLEV BARTHELMES
Institut für Binnenfischerei
Müggelseedamm 310
Berlin-Friedrichshagen
DDR-1162

Limnologica (Berlin) 19 (1), 61—70, Juli 1988

Technische Universität Dresden, Sektion Wasserwesen, Bereich Hydrobiologie, Dresden, DDR

An Acoustic Fish Stock Assessment in the Bautzen Reservoir

By Heinz Schultz

With 7 Figures and 2 Tables

Key words: Hypertrophic water; Hydroacoustics; Target strength; Stock size; Perch; Roach; Predatory fish; Gillnet.

Abstract

An acoustic survey using a pulse counting system was carried out on 55 transects at daytime in June 1982 in the hypertrophic Bautzen reservoir situated in the southeastern part of the GDR. Stock size originated from the survey had to be corrected by a factor of 4.0, as the boat's engine caused an essential escapement of the fish indicated in a specific experiment. Total stock of native species was estimated to 3800 individuals and 434 kg per ha. The perch (*Perca fluviatilis*) was by far the dominant species. The share of predatory fishes (pike-perch, *Stizostedion lucioperca*; pike, *Esox lucius*; and about 8% of the perch stock) amounted to 12%. Stock density estimated lies in the range of other waters corresponding in their trophic state. The length and age compositions of principal species are given. Fish sampling was done by gillnetting.

1. Introduction

Fish stock assessments in the Bautzen reservoir are performed in the frame of the long-term field experiment in biomanipulation (cf. Benndorf et al. 1988). Besides other assessment procedures echo sounding is used as a basic approach to stock size estimation in the reservoir. This paper deals with the first acoustic stock assessment in this water and in fresh water of GDR, too. The survey was accomplished from 7—9 June 1982. The characteristics of the reservoir situated in the southeastern part of the GDR are given by Benndorf et al. (1988). A chart of water depth resulted of the acoustic survey in question is presented in Fig. 4a.

2. Echo sounding equipment

As basic device an ELAC echo sounder of type LAZ 51 was used. The technical data of the sounder are

Frequency	200 kHz
Input voltage	24 V
Pulse length	0.3 ms
Pulse sequence in the 20 m recording range	575/min
Paper speed	21.7 or 43.4 mm/min
Transducer	Transmitting and receiving; 12 cm in diameter
Beam angle	13° (total)

Time varied gain available.

For the fish stock assessment a pulse counting system was added which consists of

a pulse generator of type TR 0331 (Hungary),

a digital counter „Polydigit" (GDR),

a coupling device with a scale for adjusting sound threshold level and a converter (24 → 220 V).

The equipment was coupled and tested by H. Bremer-Rothbarth and E. Götze of the hydroacoustic team of the Institute of Marine Fisheries, Rostock.

3. Calibration of the echo sounder

3.1 Threshold sound level

The sound level has to settle at a value that allows an avoidance of fish below 10 cm and other smaller targets. The fish has to record only from that size onwards, because smaller fish could not be fished in default of gillnets of suitable mesh size. The corresponding threshold level was established by determining the minimum power value needed to record different fish sizes and serveral species fixed in the acoustic axis of the sound beam. The oscilloscope measurements were made for different distances of the target to the sound source. First measurements of the threshold level in the frame of this stock assessment were executed by E. Götze and H. Bremer-Rothbarth.

3.2 Beam angle

The diameter of the sound beam was determined by measuring the distance at which a fixed target was recorded when the transducer is crossing. The angle to the acoustic axis (Fig. 1) is given then by

$$\text{tangens}\ \alpha = \frac{\text{radius}}{\text{distance}}\,.$$

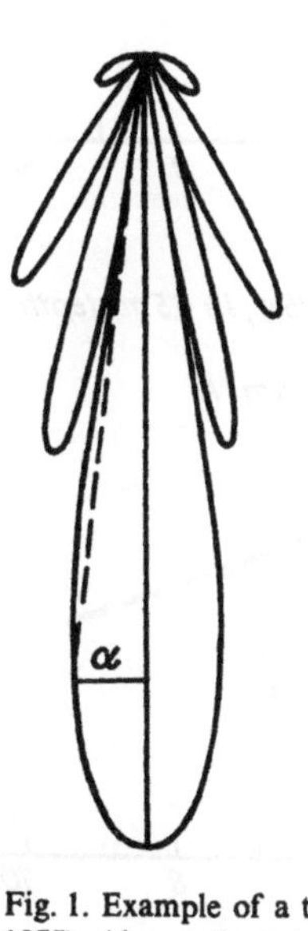

Fig. 1. Example of a transducer directivity pattern (from Saville 1977) Almost the total energy is emitted within a 6.5° (α) sector around the acoustic axis in case of the transducer used.

3.3 Time varied gain (TVG)

Into the echo sounder a TVG is installed. In case that the time or the distance dependent loss of the sound transmitted, respectively, is not fully compensated by the TVG, one can expect that the loss in the depth range above 10 m in which the fish occurred is small. Measurements of the beam angle made in 3.2 and 6.2 m depth indicated only a small difference in the mean values of both depths. This difference was not staistically manifested. In the stock density calculations there was no correction applied.

4. Target strength measurements

Target strengths used in the stock assessment were determined by counting the number of traces per unit distance on the echogram and the number of pulses automatically counted in an intervall. By this means 20% of total number of signals recorded on the survey could be taken into account. The majority of the traces could not be identified as a single fish.
Besides the overall gain and TVG characteristics of the echo sounder and the speed of the boat, target strength is dependent on the depth in which the fish is recorded, the position of the fish in the acoustic circular field, and the size of the fish. Target strength in a distinct depth shows a large variation, and the correlation between depth and target strength is only slightly manifested. Therefore, mean target strength by survey transects resp. part of transects and areas were taken for stock calculation. Mean target strength counted on the echogram was 3.6 signals per fish, it differed between 2.6 and 7.5 for areas.

In a second approach target strength was measured at a stationary single fish at different depths. Fig. 2 shows the dependence of signals per target on the speed of the boat. It proved that in many trials of passing over the fish was not hitted by the acoustic axis. Therefore, a curve was drawn through the maxima. Results of measurements made in dependence on length of principal species are summarized in Fig. 3. Target strength of perch in 6 m depth seems to be quite well, but in 4 m depth it is rather bad established. Taken into account the results given in Fig. 3, suggestions on the target strength of some subordinate species and the mean length of species, mean number of signals per fish of the total stock amounts to 5.7. On the assumption that the fish will be distributed randomly in a circular field with respect to the boat's course, the mean path length of a fish is given by equ. (1):

$$\bar{P} = \frac{\pi D}{4} \qquad (1)$$

where: $\bar{P}$ — path length; D — diameter of the field (SAVILLE 1977).

Hence the mean "on survey" target strength amounts to 4.5 signals.

The lower "on survey" target strength obtained from echograms could be explained by a possibly shorter path length of the fish than estimated and/or by a larger tilt angle than normally, both caused by an essential avoidance behaviour established in a specific experiment.

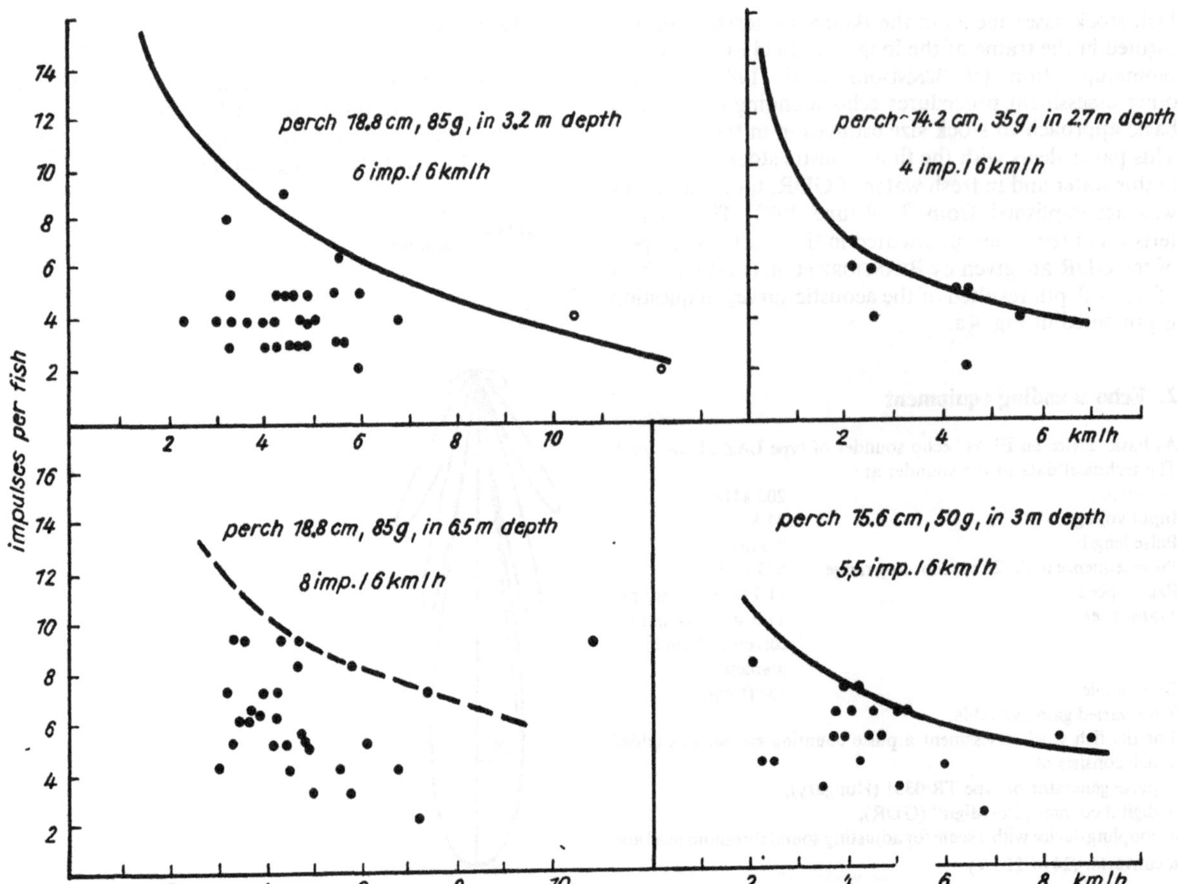

Fig. 2. Target strength measurements at stationary single perches recorded with different speed. A curve drawing through the maxima probably represents the dependence of target strength on the boat's speed.

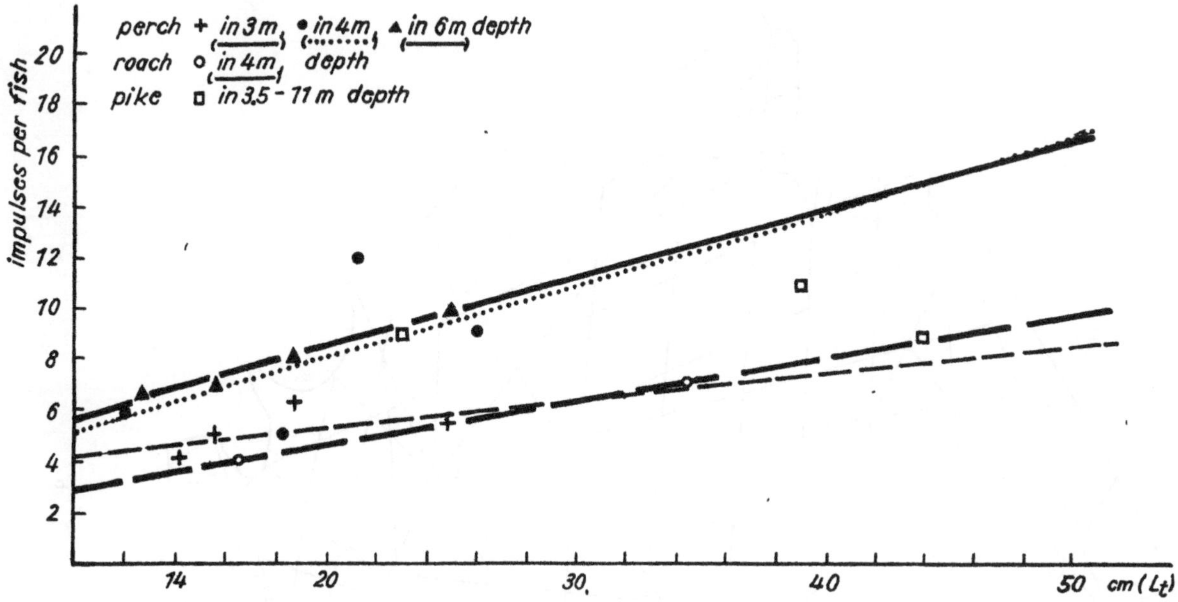

Fig. 3. Dependence of target strength on fish length (recorded with a speed of 5—6 km/h).

NAKKEN & OLSEN (1977) have shown that changes in aspect of only 7° can reduce the target strength essentially. Therefore, it was finally decided to perform the stock size calculation on the basis of target strength obtained from echograms.

5. Acoustic survey performance

The survey was carried out at the season, when the overwhelming majority of the fish does not live at the bottom. The survey grid is shown in Fig. 4b. Decisive for the choice of this survey grid was the necessity to find suitable terrestric marks for navigation at a straight line. 55 Transects were accomplished covering in total 0.7% of the reservoir area at the mean fish distribution depth of 3 m or 1.7% at the mean depth of the reservoir of about 7 m. All transects were performed between 9 a.m. and 6.30 p.m. in order to avoid miscalculations which could be caused due to diurnal fish migrations in the early morning and in the evening from and to the shore. These migrations are well known (HASSLER & VILLEMONTE 1953, THORPE 1977), and they were also observed in the Bautzen reservoir. The transducer was pulled alongside of the boat with a mean speed of 4.2—6.7 km/h per transect. The soundings were accomplished within a 20 m scale of the sounder.

6. Procedure of total stock size calculation

The number of signals was registered one each half minute, and later on, was put together to one minute intervalls. The estimation was made on area basis. Because of the increasing area of the sound beam with depth, mean depth of fish occurrence in the intervalls was ascertained for the density calculations. A further preparatory work delt with a control of the values recorded regarding overestimation by noise from non-biological sources (such as air bubbles or noise from power supply) and underestimation in the case of occurrence of fish traces outside the counting range. To this a first correction was already accomplished during the survey.

Mean density $\bar{x}$ per m² and intervall was calculated by means of equ. (2):

$$\bar{x} = \frac{I}{i \cdot 2(\tan\alpha \quad R)\, l} \qquad (2)$$

where: I — number of signals recorded; i — "on survey" target strength; α — beam angle of the acoustic axis; R — mean distance (m) of the targets to the transducer, l — distance (m) covered by the boat.

Isolines of stock densities were drawn, and total number of fish in the reservoir and in subareas was calculated by means of computer planimetry. Unfortunately, it proved to be necessary to correct the number of fish calculated for escapement behaviour caused by the outboard motor used. According to specific experiments escapement behaviour contributed to a great bias in the stock assessment. The mean correction factor derived from the experiments and applied to the original stock size estimation was 4.0. The experiments were executed on two boats. One boat anchored at several positions in the reservoir at a water depth of 5 m where fish was more or less steadily present near and up to 1.5 m above the bottom (in fewer cases 2—4 m above) sheltered the echo sounding equipment. In many trials the other boat driven by the outboard motor passed near the transducer alongside of the first boat with the corresponding speed. In almost all cases an escapement of fish could be distinctly established, but a large variation of the rate of reduction of signals recorded was also observed. In most cases the reduction of the echo abundances amounted to 60—90% and on average to about 75%.

7. Estimation of fish number per species

The number of individuals of fish species was estimated as follows:

(1) The number of fish of those species not catched by the test fishing for lack of mesh sizes above 80 mm were calculated first. These non-native species (bighead, *Aristichthys nobilis*; silver carp, *Hypopthalmichthys molitrix*; grass carp, *Ctenopharyngodon idella*) were esti-

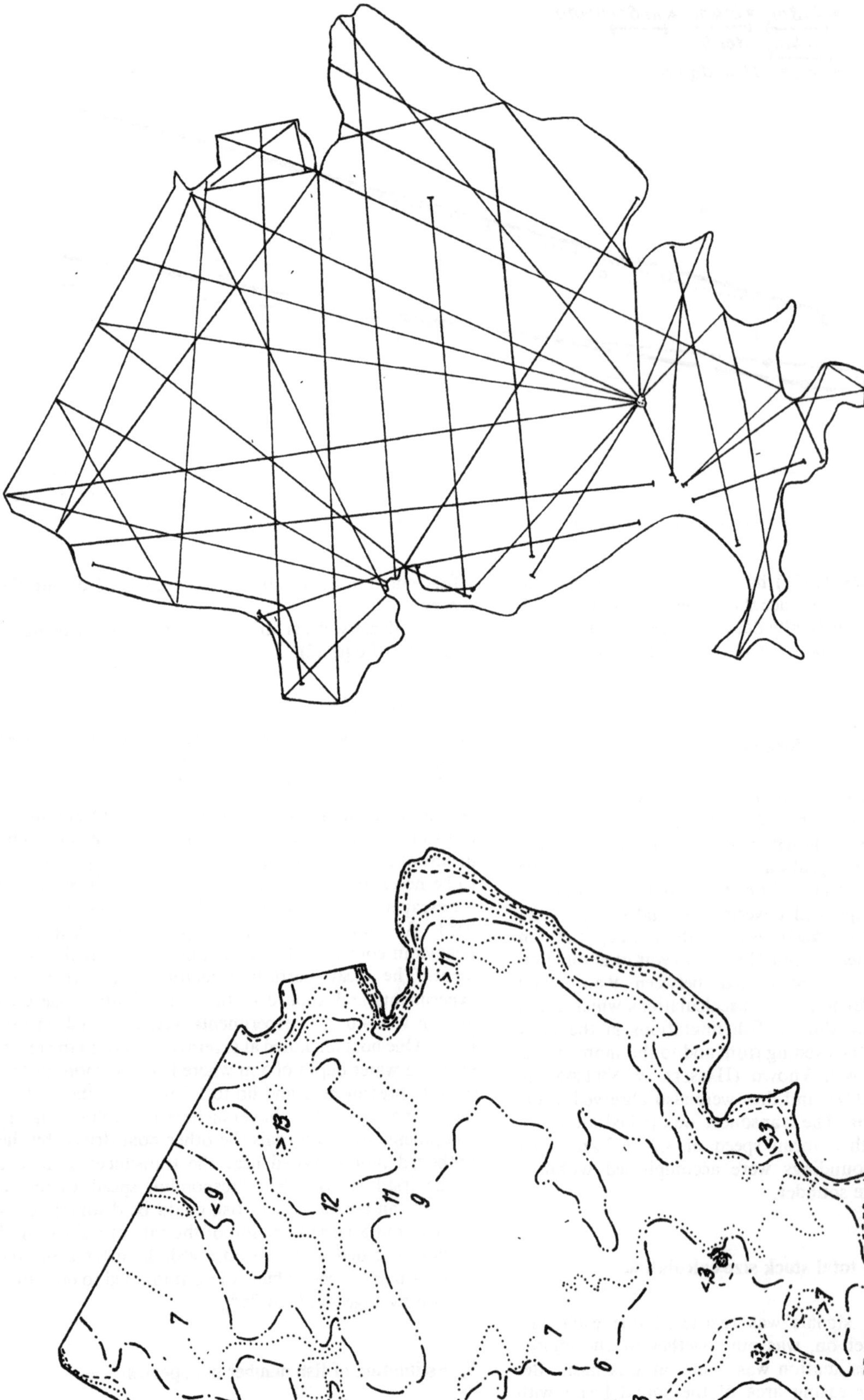

b

a

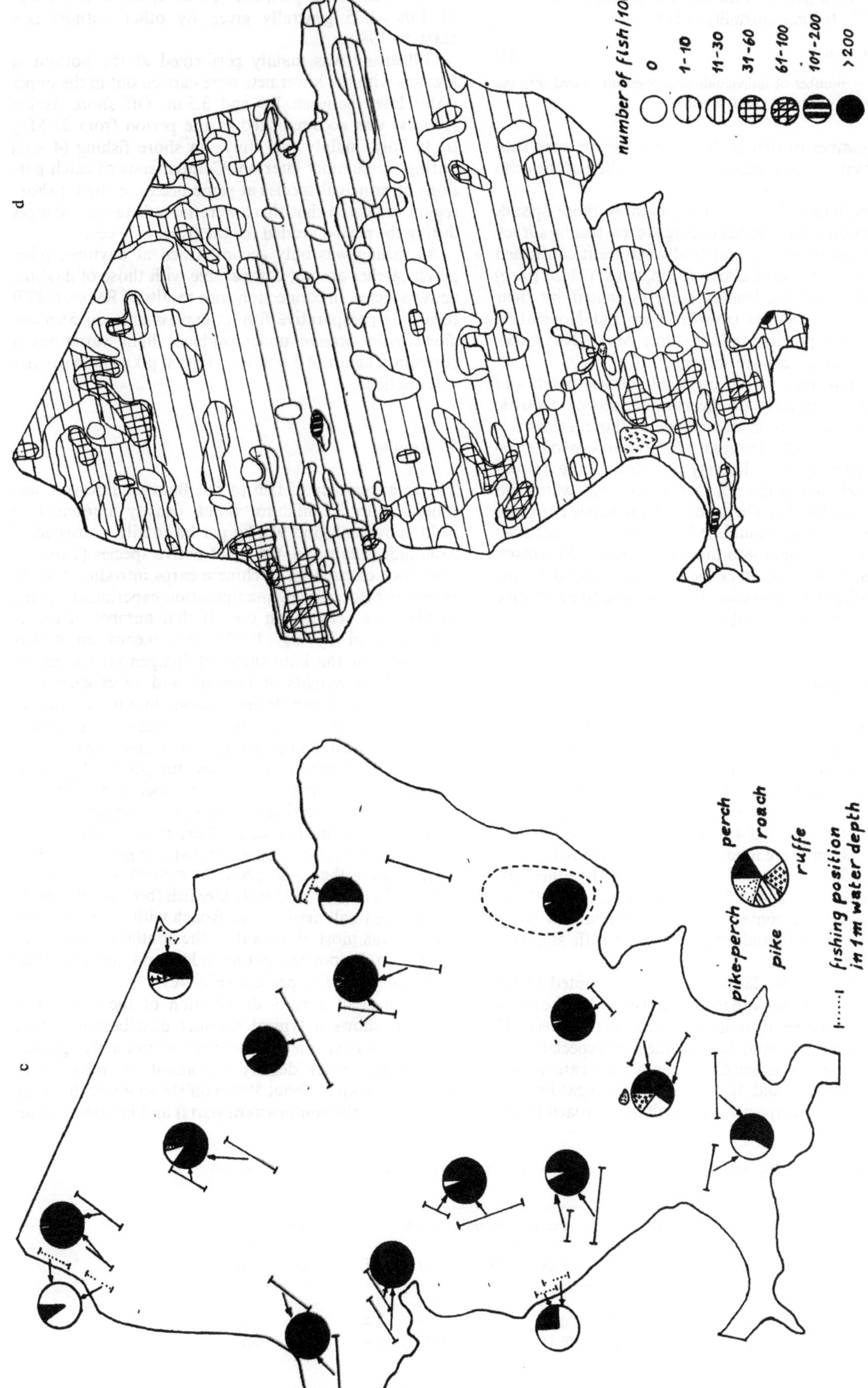

Fig. 4. The Bautzen reservoir.
a — chart of water depth at normal water level, b — acoustic transects performed 7—9 June 1982, c — shares of principal species in the gillnet catches during the survey; d — fish stock density distribution. Densities in the figure are not corrected for escapement of fish (further explanation see text)

mated on the basis of the number introduced into the reservoir and a guess on the natural mortality ($M = 0.1 - 0.05\ a^{-1}$; fishing mortality is zero):

$$N_{k,n+1} = N_{k,n} \cdot e^{-Z} \quad (3)$$

where: N_k — number of individuals of species introduced into the reservoir; n — year; Z — total mortality

(2) The number of fish of these three species were subtracted from the total number of fish resulted of the echo survey.
(3) On the basis of the resulting number those species which portion in the catches during the test fishing survey were believed to be not statistically sufficient confirmed [pike, pike-perch, perch above 30 cm, tench (*Tinca tinca*) and bream (*Abramis brama*)] were estimated by their portions in the catches between May and November. In Table 1 their portions in June—July, May—November and for the total year are shown.
(4) The abundance of the principal species, perch and roach, was ascertained on the base of subarea acoustic results and the share in the catches of those areas.
(5) Test fishing indicated that ruffe (*Gymnocephalus cernua*) was distributed exclusively very close to the bottom and not included in the echo counting range which did not sweep the bottom. On the other hand single fish close to a muddy bottom should be impossible to detect even with a white line built into the echo sounder (MATHISEN 1980). Therefore, the stock of ruffe was added to the stock of other native species corresponding to its relative share in the catches of subareas.

8. Fish sampling

Test fishing for identification of fish species and size was done by gillnetting. Gillnets of eight mesh sizes (13, 18, 24, 28, 32, 45, 50 and 60 mm; control fishing with 80 mm mesh size outside the survey had always given zero catches) and length of about 45 m (with the exception of the 24 and 45 mm nets which had a length of 23 resp. 34 m) were used. The height of the nets of the 24, 28 and 32 mm mesh size were 4—5 m, the other nets were between 1.6 and 2.5 m. Nominal catches of the different nets were converted to catch per unit of effort [standard time and standard net area (for ruffe standard net length)].

The mesh sizes 24, 28 and 32 mm contributed to the majority of the catches and to almost the total of the perch catches. The larger distance in mesh size between 32 and 45 mm which could have influenced especially the catches of roach was tided over by a high catchability efficiency of the 28 and 32 mm mesh sizes regarding the length range of this species. The girth of the roach fished by these mesh sizes amounted up to a mean value of 1.6 times the mesh perimeter on comparison to a range of 1.08—1.35 generally given by other authors (see HAMLEY 1980).

Gillnetting was mainly performed at the bottom, a few sets with the lower nets were carried out in the upper water layer between 1.0 and 3.5 m. Off shore fishing (13 sets) was accomplished in the period from 27 May to 15 July, mainly to 1 July, on shore fishing (4 sets) during the summer later on. Comparisons of catch portions of principal species at some places in the off shore region indicated that there were no substantial changes during the period needed for the fishery survey.

As fishing was only accomplished at daytime, pike-perch catches are not comparable with those of daytime active species. In connection with results of RYDER (1977) regarding comparative fishing experiments on *Stizostedion vitreum vitreum* under ambient light variations a correction factor of 3.0 to the original pike-perch catches was applied.

9. Results

The total number of fish in the Bautzen reservoir was estimated to 2.2 millions. Stock density amounted to 3870 individuals and 920 kg per ha totally, consisted of 3800 individuals and 434 kg of native species (Table 2). The stock calculation of chinese carps introduced in the reservoir before the biomanipulation experiment started in 1977 is a very rough one. If that number of fish is overestimated by e.g. 100%, this means an underestimation of the individuals of indigenous species by 2.5%. Mean weights of bighead and silver carp were derived from growth determinations in natural lakes of the GDR (KOZIANOWSKI 1981) and that of grass carp in connection with data given by JÄHNICHEN (1981).

The most abundant species was the perch followed by the roach. The third species by number, the ruffe, only occasionally occurred in the catches. The highest portion of ruffe was attained in the southern part of the reservoir. Regarding the ruffe stock estimated one has to take into consideration that this species assessment comprehends only the 3-summer olds and older fish (because of missing of suitable small mesh sizes). Roach within the off shore regions was most abundant in the southern part of the water, too. Its portion in the middle and northern areas was regularly a few percent (Fig. 4c).

The chart of density distribution of the total stock (Fig. 4d) shows a typical summer distribution picture with generally smaller abundances in and above greater water depth. Stock density was about six times higher in the region up to about 500 m off the shore (without the exception of the southeastern parts) and in bights. Main

Table 1 Relative share (%) and number (N in 10^3) of fish species occurring in small quantities in the catches of 1982

Month	Pike		Pikeperch[1]		Perch, >30 cm		Tench		Bream	
	(%)	(N)	(%)	(N)	(%)	(N)	(%)	(N)	(%)	(N)
VI—VII	0.10	2.1	0.15	3.1	0.20	4.2	0.35	7.3	—	—
V—IX	0.12	2.5	0.15	3.1	0.15	3.1	0.21	4.4	0.18	3.8
Total year (III—IX)	0.41	8.5	0.20	4.2	0.15	3.1	0.18	3.8	0.15	3.1

[1] Not corrected to night-catches

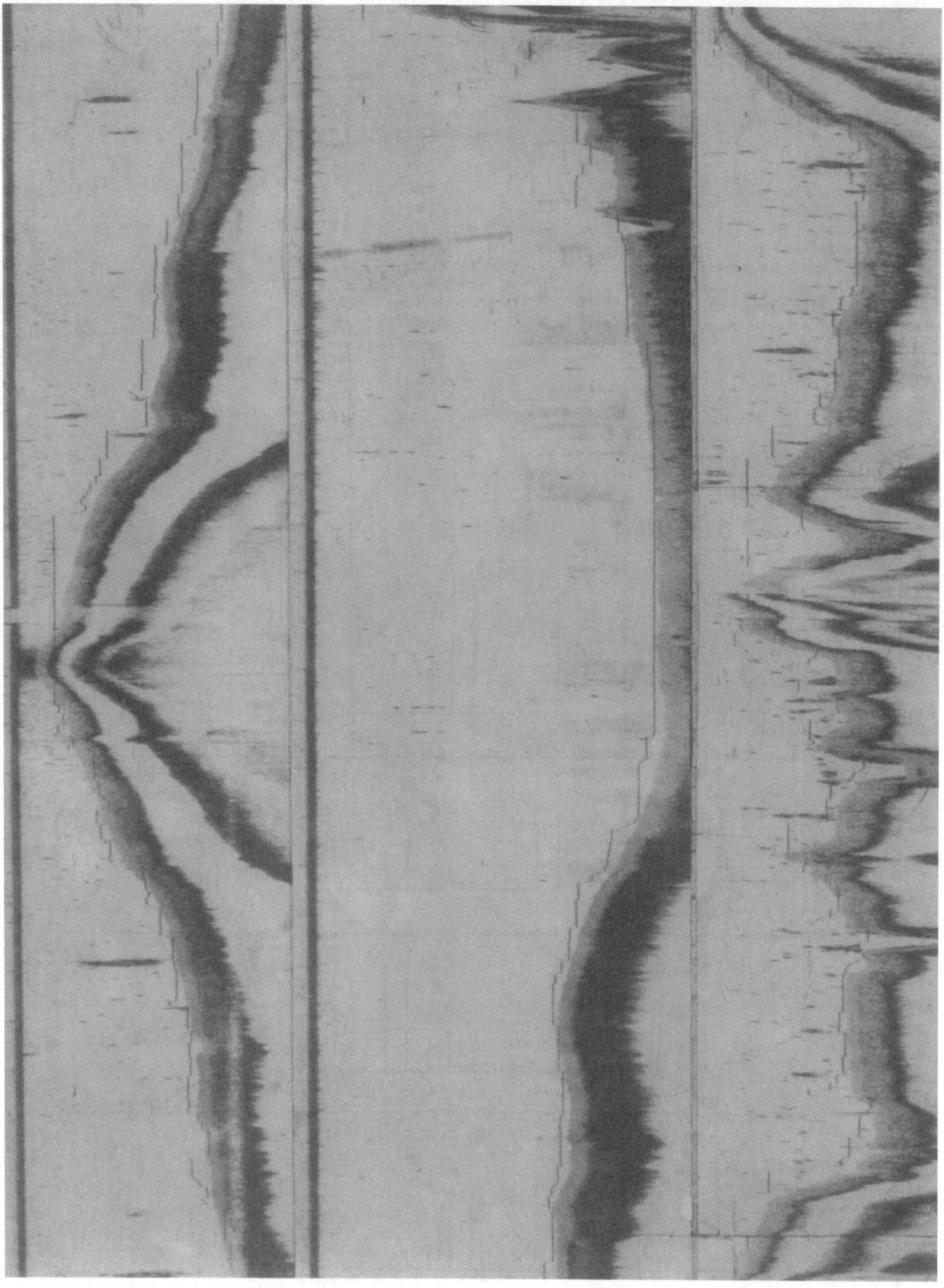

Fig. 5. Echograms of the Bautzen reservoir.

Upper: Parts ot two transects in the northern area of the reservoir Middle Continuation of the upper right hand transect towards the eastern shore. At the foot Three short transects in the southern part of the reservoir Maximum water depth in the upper picture is 8 m, in the middle picture 12 m. Mean water depth in the third picture is 5—6 m. The dark line above the lake bottom marks the bottom limit of the counting range (the upper limit scarcely or not visible). Lake bottom is recorded with "grey line" switched on

Table 2. Stock density and mean weight per species in the Bautzen reservoir in June 1982

	Number per ha	Mean weight (g)	kg per ha
Perch (*Perca fluviatilis*)	2862	99	283.4
Perch, >30 cm	6	750	4.1
Roach (*Rutilus rutilus*)	611	164	100.2
Ruffe (*Gymnocephalus cernua*)	218	15	3.2
Pike-perch (*Stizostedion lucioperca*)	17	1080	17.9
Pike (*Esox lucius*)	4	2300	10.4
Tench (*Tinca tinca*)	8	910	7.1
Bream (*Abramis brama*)	7	100	0.7
Rainbow trout (*Salmo gardneri*)[1]	71	100	7.1
Grass carp (*Ctenopharyngodon idella*)[2]	34	5100	175.5
Bighead (*Aristichthys nobilis*)[2]	25	11000	274.3
Silver carp (*Hypophthalmichthys molitrix*)[2]	7	5300	35.9
Native species	3804	116	434.1
Chinese carps[2]	66	7370	485.7
Total	3870	239	919.8

[1] Stocked one week after the acoustic survey
[2] Rough estimate

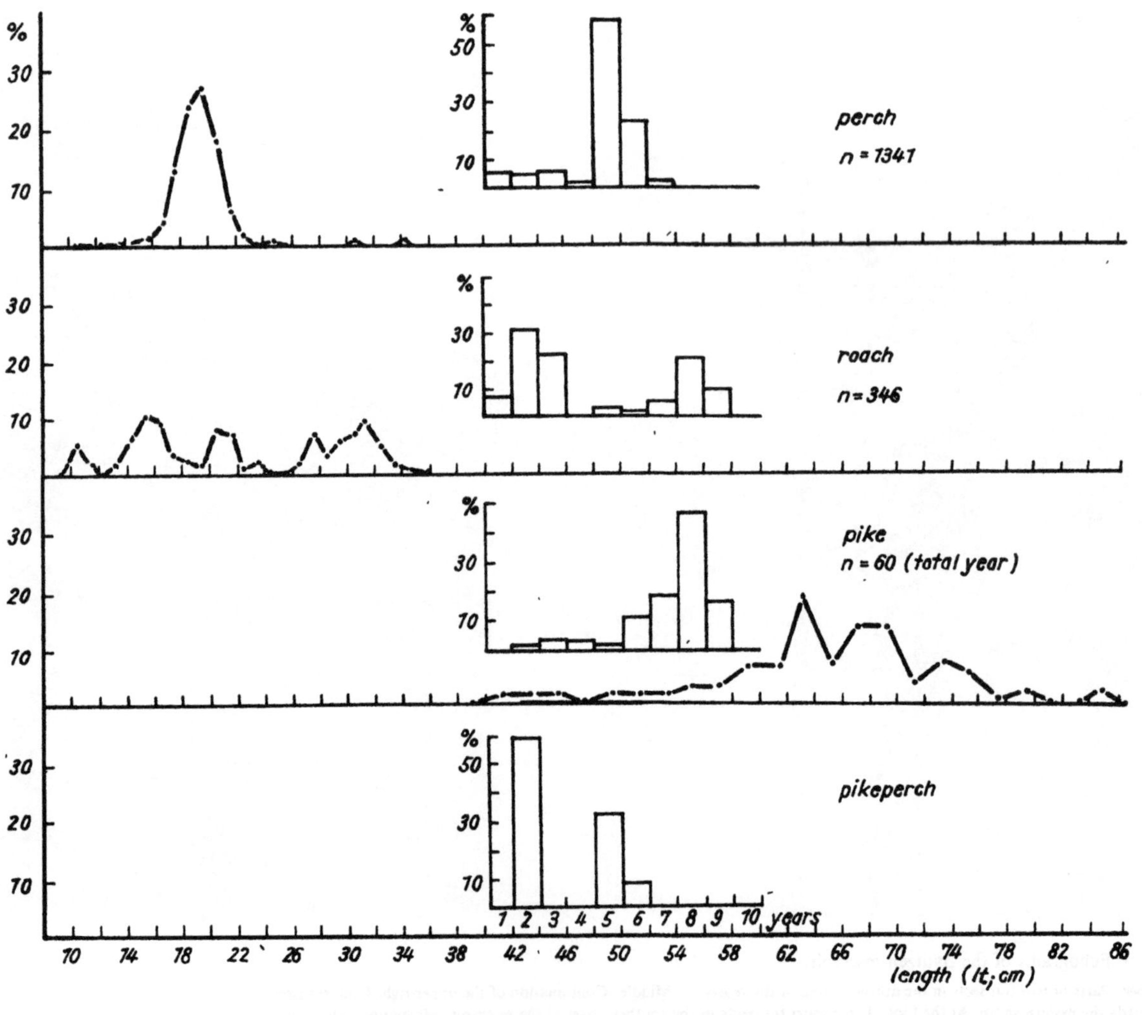

Fig 6. Length and age composition of principal species in the Bautzen reservoir in June—July 1982.

(Age composition of pike-perch was estimated from number of stockings and catches, one-year olds not included, the length-frequency distribution is not shown for this species because of non-representative amounts in the catches of 1982).

vertical distribution range was in depth of 1—5 m. Mean fish depth was 3 m, fish depth of 4 m was most frequently observed, and maximum depth at which fish was recorded was 11 m. Fish stock recorded appeared as single individuals and small shoals (Fig. 5).

The length and age composition of principal species is demonstrated in Fig. 6. In the perch stock the essential lengths ranged from 17—20 cm and the year-class 1977 dominated by about 60%. Roach existed in the catches in a length range from 10—34 cm with dominating length groups of 14—16, 20—21 and 27—32 cm representing the year-classes 1980, 1979 and 1977—1972. The year-class 1974 formed the outstanding one amongst the adult part of the population. In the pike stock the year-class 1974 dominated by about 50%. Although the number of pike investigated is rather small, the age composition is representative, as indicated by sampling in the foregoing year. The pike-perch population consisted in summer 1982 of the year-classes 1981, 1980 and 1977 introduced in the reservoir. Moreover, a smaller percentage of fish was determined as year-class 1976.

10. Discussion

A comparison of the result of density estimation of native species in the Bautzen reservoir in 1982 with other waters shows that stock density obtained is in the range given for other waters, if stock density is correlated to total phosphorus of the water. According to the correlation given for Canadian lakes by HANSON & LEGETT (1982) stock density of the Bautzen reservoir lies at the upper limit (Fig. 7). To this it must be mentioned that Middle European and Canadian waters are not fully comparable because of the somewhat shorter vegetation period of Canadian regions.

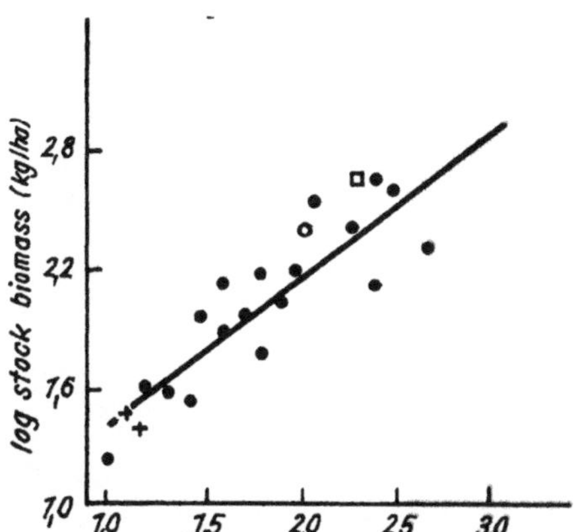

Fig. 7. Relation between total phosphorus in the waters and fish stock biomass

[from HANSON & LEGETT (1982), four waters added: Bautzen reservoir (□) — total phosphorus value from BENNDORF (1987); Saidenbach reservoir (●) — total phosphorus value made available by Laboratory of Hydrobiology Neunzehnhain, fish stock — own data unpublished, two Swedish forest lakes (+) — from NYBERG (1979)]

According to a correlation between gross primary production (g C/m² · a) or trophic state and fish yield derived from 42 lakes, resèrvoirs and coastal waters by BULON & VINBERG (1981, cited in BARTHELMES 1984) the eventual yield of the Bautzen reservoir lies somewhat beneath the regression line assuming a relation factor of sustainable yield to stock size of 6 to 5 and a hypertrophic state of the water.

The rather high fish biomass in the Bautzen reservoir seems to be supported by a moderate portion of predatory fish stock biomass (perch above 30 cm and about 7% of perches in the length range of 10—24 cm, pike-perch and pike) which amounted to 12%. Taken into consideration a correlation of total yield and the share of predatory fish in the catches described by BONAR (1977) for 43 Polish pike-perch lakes, the portion of predatory stock biomass could mean that total stock biomass of native species (not the number of fishes) was in 1982 near the maximum (on the ascending part of the curve).

Although the result of the stock assessment seems to be in a rather sufficient accordance with stock sizes in corresponding waters, in future acoustic assessments further investigations have to be made on the rate of escapement of fish in reaction to noise of the engine, or procedures have to be applied avoiding the escapement of fish, as shying of the fish could contribute to the heaviest bias in the assessment of shallow water stocks. In a recent acoustic survey comparisons were made in sounding transects conducted by an electro-motor and afterwards by a benzine-motor. The results of this comparison are not yet available.

An other aspect to be considered is the missing of the 0-group fish assessment (in case of ruffe the 0- and I-group assesment). In order to ensure a rough estimate of that age-group(s), our efforts are directed to calculations of stock sizes backwards to the 0-group on the base of consumption and feeding selection of predators. In preliminary trials numbers per age-group of perch and roach were calculated for the period 1981 to 1977 by a „multispecies“ model in which several simplifications had to be made indeed. The total number of the perch and roach stock estimated in such a way are demonstrated by BENNDORF et al. (1988).

11. Summary

An acoustic survey using an impulse counting system was carried out on 55 transects at daytime from 7 to 9 June 1982. The fish stock recorded appeared as single fish and small shoals in the mid-water. The overall mean of "on survey" target strength amounted to 3.6 signals. Escapement behaviour of the fish caused by the boat's engine contributed to a great bias in the stock assessment. According to a specific experiment on the avoidance behaviour in the fish a correction factor of 4.0 to the original stock estimation was applied. The total stock of native species was calculated to 3800 individuals and 434 kg per ha. Perch was by far the dominant species followed by roach. The share of predatory fishes (pike-perch, pike, and about 8% of the perch stock) amounted to 12%. A rough estimation of the stocks of non-native bighead, silver carp, and grass carp, which were introduced into the reservoir before the beginning of the biomanipulation experiment in 1977, results in a value of 66 individuals and 486 kg per ha and might be considered as a maximum figure. An overestimation of e.g. 100% would result in an underestimation of the native stock by only 2.5%. In comparison with other waters, stock density of native species estimated lies in the upper range given by correlation between total phosphorus and stock density (from HANSON & LEGETT 1982) resp. seems to be somewhat beneath the regression line resulting of primary gross production and fish yield (from BULON & VINBERG 1981; BARTHELMES 1984).

Fish sampling was done by gillnetting. The length and age composition of perch, roach, pike, and pike-perch were described.

Zusammenfassung

Es wird eine Fischbestandsschätzung an Hand eines hydroakustischen Verfahrens (Impulszählsystem) in der hypertrophen Talsperre

Bautzen im Rahmen eines Langzeitexperimentes zur Biomanipulation vorgestellt. Abundanzen, Biomassen, Größen- und Altersklassenzusammensetzung werden mitgeteilt. Der Raubfischanteil beträgt 12%

12. Acknowledgements

I am obliged to the Oberflußmeisterei Karl-Marx-Stadt for placing at our disposal the ELAC echo sounder, and I thank Dipl.-Phys. E. GÖTZE and Dipl.-Ing. H. BREMER-ROTHBARTH of the Institut für Hochseefischerei und Fischverarbeitung, Rostock, for their helpful instructions on acoustic theory, valuable advices on calibration methods and acquisition of attachments as well as for the coupling and testing of the echo counting equipment.

References

BARTHELMES, D. (1984): Neue Gesichtspunkte zur fischereilichen Seenklassifizierung und daraus ableitbare Möglichkeiten der Ertragssteigerung. Fortschr. Fischereiwiss. 3, 133—144.

BENNDORF, J. (1987): Food web manipulation without nutrient control: A useful strategy in lake restoration? Schweiz. Z. Hydrol. 49 (2), 237—248.

— SCHULTZ, H., BENNDORF, A., UNGER, R., PENZ, E., KNESCHKE, H., KOSSATZ, K., DUMKE, R., HORNIG, U., KRUSPE, R., & REICHEL, S. (1988). Food-web manipulation by enhancement of piscivorous fish stocks: Long-term effects in the hypertrophic Bautzen reservoir. *Limnologica* (Berlin) 19 (1), 97—110.

BONAR, A. (1977): Relation between exploitation, yield and community structure in Polish pikeperch (*Stizostedion lucioperca*) lakes, 1966—71. J. Fish. Res. Bd. Can. 34 (10), 1576—1580.

HAMLEY, J. M. (1980): Sampling with gillnets. In: BACKIEL, T., & R. L. WELLCOMME (eds.), Guidelines for sampling fish in inland waters. FAO, EIFAC Techn. Paper Nr. 33, 37—53.

HANSON, J. M., & LEGETT, W. C. (1982): Empirical prediction of fish biomass and yield. Can. J. Fish. Aquat. Sci. 39 (2), 257—263.

HASSLER, H. D., & VILLEMONTE, J. R. (1953): Observations of the daily movements of fishes. Science 118, 321.

JÄHNICHEN, H. (1981): Produktion pflanzenfressender Cypriniden. In: W. STEFFENS, Industrielle Fischproduktion. 2. Aufl. Berlin, S. 135—154.

KOZIANOWSKI, A. (1981): Der Stand und die Perspektiven der Produktion pflanzenfressender Fische in der Binnenfischerei der DDR. Z. Binnenfisch. 28, 355—362.

MATHISEN, O. A. (1980). Acoustic stock assessment, In: BACKIEL, T., & R. L. WELLCOMME, Guidelines for sampling fish in inland waters. FAO, EIFAC Techn. Paper Nr. 33, 115—141.

NAKKEN, O., & OLSEN, K. (1977): Target strength measurements of fish. Rapp. et Proces-Verbaux des Reun. 170, 52—69

NYBERG, P. (1979): Production and food consumption of perch, *Perca fluviatilis* L., in two Swedish forest lakes. Inst. Freshwater Res., Drottningholm, Rep. Nr. 58, 140—157.

RYDER, R. A. (1977): Effect of ambient light variations on behavior of yearlings, subadult, and adult walleye (*Stizostedion vitreum vitreum*). J. Fish. Res. Bd Can. 34 (10), 1481—1491.

SAVILLE, A. (1977): Survey methods of appraising fishery resources. 3. Acoustic surveys. FAO, Fish. Techn. Paper Nr. 171, 20—38.

THORPE, J. (1977): Synopsis of biological data on the perch, *Perca fluviatilis* LINNAEUS, 1758 and *Perca flavescens* MITCHILL, 1814. FAO, Fish Synopsis Nr. 113.

Received January 23, 1987

Address of the author:

Dr. rer. nat. HEINZ SCHULTZ
Technische Universität Dresden
Sektion Wasserwesen
Bereich Hydrobiologie
Mommsenstr. 13
Dresden
DDR-8027

Part II

Studies of Integral Food-webs

Limnologica (Berlin) 19 (1), 73—79, Juli 1988

Department of Zoology, University of Washington, Seattle, U.S.A.

Unplanned Biomanipulation in Lake Washington

By W. T. Edmondson & Sally E. B. Abella

With 4 Figures and 2 Tables

Key words: Biomanipulation; Keystone predator; Reproductive success; Manipulation of chemical conditions, *Daphnia*; Smelt

Abstract

Because *Daphnia* is able to increase the clarity of lakes by consuming small algae, it is an effective tool in biomanipulation of eutrophied lakes. Lake Washington became much more transparent after an unexpected outbreak of *Daphnia* in 1976 than it had been before. The data strongly suggest that *Daphnia* had previously been suppressed by two conditions, a relatively abundant population of the selective predator *Neomysis* and an abundance of the alga *Oscillatoria* which interferes with the feeding mechanism of *Daphnia*. Diversion of sewage effluent clearly was responsible for the decrease in *Oscillatoria*. The reduction in *Neomysis* was correlated with an increased abundance of the longfin smelt, a fish that feeds selectively on *Neomysis*. The increase of smelt was probably facilitated by improvements in the physical conditions of the Cedar River that promoted spawning and survival. There is also a possibility that growth of the smelt population was aided by a decrease in the predatory Northern Squawfish. This experience suggests that desirable predators may be increased by manipulating the physical conditions that increase reproductive success. For a manipulation of physical conditions to be successful, one must take account of the biological relationships of the species present.

Introduction

The concept of biomanipulation has been developed to include a group of techniques for the management or protection of lakes against eutrophication as an alternative to techniques that involve manipulating the chemical control system. Techniques of biomanipulation change the interactions among species in the lake in such a way as to improve the condition of the lake as perceived by people (Shapiro et al. 1975, Shapiro 1979, Shapiro 1980). Any kind of interaction can be considered for use in a program of biomanipulation. While exploitative or interference competition evidently can have pronounced effects in some kinds of communities (Schoener 1983), often a sudden, prolonged change in the abundance of a given species has been preceded by a change in the abundance of a predator. In many cases a cause-and-effect relationship has been demonstrated experimentally, or shown to be very likely on general principles. Sometimes apparently competitive interactions have a predatory component for some of the stages of life history, as with interactions between species of flour beetles or copepods of different adult size. Indeed, the concept of keystone predator is well established, meaning that some predators are able to eliminate certain prey species, and their effect is out of proportion to their abundance (Paine 1969, 1980).

Different ways of using keystone predators have been successful in lakes. In some cases, essentially the entire fish population has been eliminated by poisoning with rotenone, and a more desirable combination of species added (Shapiro & Wright 1984). Sometimes a predatory species is added to an existing community in a lake not previously inhabited by it. For example, the introduction of the crustacean *Mysis relicta* has usually been followed by large changes in the zooplankton, often by a reduction in abundance or disappearance of *Daphnia* (Rieman & Falter 1981, Langeland 1981, Morgan 1982). *Daphnia* is also subject to elimination by some planktivorous fish. In Lake Michigan, predation by introduced salmonid fish was followed by reduction in the planktivorous alewives and an increase in their *Daphnia* prey (Evans & Jude 1986, Scavia et al. 1986). In addition to direct effects on the abundance of prey species, there may be important indirect effects on the supply of nutrients to the phytoplankton (Wright & Shapiro 1984).

Poisoning with rotenone is not practicable in very large lakes, and in any case may damage desirable populations. In other cases, introduction of a potential keystone predator may fail. A less disruptive approach is to find ways to promote the success of desirable species that are already part of the community by increasing their rate of reproduction or survival and by decreasing their death rate. This would be aided by reducing the rate of reproduction and survival of undesirable species. Thus we would encourage the lake to solve its own problems.

As mentioned above, one of the goals of a biomanipulation may be to increase the growth of *Daphnia*. Populations of relatively modest size are able to clear the water of small phytoplankton that strongly affect the transparency as measured by the Secchi disc (Lampert et al. 1986). Large colonies of some kinds of Cyanobacteria may remain; however, the water looks crumby but much clearer and generally more acceptable to the public (Edmondson 1980). Much of the literature treats the Cyanobacteria as a rather uniform group, but different genera behave differently and have different effects on zooplankton. In particular, certain species of *Oscillatoria* are of special significance to *Daphnia* as will be shown below. It will be worthwhile to improve our understanding of the conditions that lead to dominance by those species rather than by *Aphanizomenon*, *Anabaena* or *Microcystis*.

Future improvements in biomanipulation will include learning more about conditions that promote the success of *Daphnia*. A valuable guide can result from studies of lakes in which the success of *Daphnia* has changed, whether by increase or decrease, without deliberate efforts to make the change. Identification of the probable causes of such changes can lead to the design of effective experiments as a basis for improved biomanipulation. Lake Washington has provided such an experience (Table 1).

Table 1 Representative data for Lake Washington and Cedar River

Lake Washington	
Area (m^2) 87,615 × 10^3	
Volume (m^3) 2,885 × 10^6	
Mean depth (m) 32.9	
Fraction renewed (y^{-1})	Mean 0.428 Range 0.318—0.591 (1962—1978)
Retention time (y)	Mean 2.34 Range 3.14—1.69 (1962—1978)
Dissolved P loading ($g \cdot m^{-2} \cdot y^{-1}$)	Mean 1.39 (1962—1965)
	Mean 0.27 (1969—1978)
Alkalinity $CaCO_3$ (l^{-1}) 35	
Total dissolved solids ($mg \cdot l^{-1}$) 75	
Cedar River	
Total annual discharge (10^6 $m^3 \cdot y^{-1}$)	Mean 602 Range 432—891 (1960—1980)
Total discharge Jan.—Feb. (10^6 m^3)	Mean 164 Range 70—249 (1960—1980)

For more information see EDMONDSON & LEHMAN (1981)

Daphnia in Lake Washington

Between 1933 and 1973, *Daphnia* was collected infrequently and usually in small numbers. In 1976, after several years of small increases, *Daphnia* suddenly dominated the summer zooplankton with a concurrent doubling of water transparency (EDMONDSON & LITT 1982). This event happened during a time when the lake was being closely examined as part of a study of the recovery of the lake from eutrophication (EDMONDSON & LEHMAN 1981). Thus we were able to look for changes in factors known to affect *Daphnia* in search for a possible explanation. There was no way to anticipate the 1976 outburst of *Daphnia*, and on the basis of information available then, impracticable to have organized *in situ* experiments. On the basis of the existing data, of general knowledge of the requirements of *Daphnia* and of special experiments, it seemed likely that *Daphnia* increased in 1976 because it was no longer suppressed by two inhibitory conditions, one or both of which had been in effect since the first detailed study (SCHEFFER & ROBINSON 1939).

The first change was a decrease in the abundance of the crustacean *Neomysis mercedis* between 1960 and 1967 to about one tenth its former value. Because of the known effectiveness of mysid predation, *Neomysis* became a prime suspect. A special study of the feeding ecology of *Neomysis* showed that it had exactly the right selective feeding properties to exert a strong control on *Daphnia* populations (MURTAUGH 1981).

The success of *Daphnia* cannot be attributed to release from predation by *Neomysis* alone. Nearly ten years elapsed between the decrease of *Neomysis* and the increase of *Daphnia*. No correlated decrease of planktivorous fish capable of controlling *Daphnia* is known. On the contrary, the sockeye salmon (*Onchorhynchus nerka*) increased during the time of decrease of *Neomysis*, and it has remained present during the increase of *Daphnia* (EGGERS et al. 1978). It eats *Daphnia*, but evidently the rate of predation is insufficient to eliminate *Daphnia*. Further, in 1977 the Washington Department of Game started a program of planting fingerling rainbow trout (*Salmo gairdneri*). These small fish feed extensively on *Daphnia*, but so far the *Daphnia* population has been of similar size each summer (EDMONDSON 1985). The explanation for the delayed increase of *Daphnia* must depend on some process other than predation.

Enough food for *Daphnia* was present (INFANTE & ABELLA 1985), but changes took place in the composition of the phytoplankton that are known to be important to *Daphnia*. Specifically the abundance of *Oscillatoria* progressively decreased until it disappeared after 1976. The inhibitory effect of *Oscillatoria* on *Daphnia* primarily by mechanical interference with feeding rather than toxicity was cited as the probable cause of the delayed success of *Daphnia* (EDMONDSON & LITT 1982). A special study of this effect was made showing that reproduction and early growth were reduced by moderate amounts of *Oscillatoria* (INFANTE & ABELLA 1985, see also DAWIDOWICZ et al. 1987). Thus, 1976 evidently was the first year in which Lake Washington did not have inhibitory abundances of *Neomysis* or *Oscillatoria* (Table 2). It should be noted

Table 2. Correlations of abundance of organisms in Lake Washington

Time	Smelt	*Neomysis*	*Oscillatoria*	*Daphnia*
1933—1954	—	+	—	—
1955—1967	—	+	+	—
1968—1975	+	—	+	—
1976—present	+	—	—	+

— scarce or infrequent
\+ abundant or consistently present
Note that *Daphnia* was abundant only when both *Oscillatoria* and *Neomysis* were scarce

that the mechanical interference produced by *Oscillatoria* is not general for all filamentous algae nor all Cyanobacteria, nor possibly for all species of *Oscillatoria*. *Daphnia* can make good use of some filamentous diatoms (INFANTE & LITT 1985, INFANTE & EDMONDSON 1985). Some blue-green colonies such as *Aphanizomenon flos-aquae* are usually too big to interfere with *Daphnia*'s feeding. Toxicity is an important but not universal feature of Cyanobacteria (NIZAN et al. 1986).

The changes in phytoplankton can be regarded as consequences of changes in the trophic state of Lake Washington, and therefore the success of *Daphnia* in part is a consequence of the reduction in nutrient input. The change in *Neomysis* must have another cause. As usual, it is reasonable to look for a change in a potential keystone predator; in this case a likely one can be found.

Predation on *Neomysis*

The only recorded change in a predator on *Neomysis* seems to be an increase in the abundance of the longfin smelt (*Spirinchus thaleichthyes*). Before 1960 it must have been either absent or below the level of detection by existing sampling programs. The first known sighting of the smelt in Lake Washington was in September 1960 by ROBERT L. DRYFOOS during a field trip by a limnology class. A large sample had been taken with a 2 m midwater trawl to demonstrate the spectacular *Neomysis* population when DRYFOOS, a member of the class, noticed several smelt. Similar hauls taken in previous years had not contained smelt. For instance, a haul taken in December 1958 by Dr. JOSEPH SHAPIRO contained approximately 3,400 *Neomysis* but no smelt. The chance observation in 1960 led to two detailed studies (DRYFOOS 1965, MOULTON 1974; see also EGGERS et al. 1978). The smelt is a specialized predator on *Neomysis*. During 1962 to 1964 about 85% of the gut contents of smelt were *Neomysis* (EGGERS et al. 1978). In 1985, smelt contained *Neomysis*, but also had *Daphnia* (D. A. BEAUCHAMP, University of Washington School of Fisheries, pers. comm.). A few other species of fish eat *Neomysis*, but not so selectively, and none is known to have increased as did the smelt; however, the data on abundance are not extensive enough to define the long term changes of all the species of fish.

Longfin smelt in Lake Washington

The anadromous longfin smelt is distributed along the west coast of North America from northern California to northern British Columbia, breeding in small coastal streams (WYDOSKI & WHITNEY 1979). The Lake Washington population is one of two known exceptions in that it passes its entire life in fresh water. Maturing at two years, with a mean standard length ranging from 105 to 125 mm, the fish enter tributaries of the lake to spawn, principally the Cedar River (MOULTON 1970).

The presence and behavior of the smelt in Lake Washington can be understood on the basis of changes in the drainage of the lake (Fig. 1). Before 1916, the lake drained from the south end through the Black River which was joined by the Cedar River, eventually entering the Duwamish River that connects to the marine waters of Puget Sound. A new outlet was formed by building a ship canal with locks and fish ladder connecting the lake with Puget Sound. The level of the lake was lowered by about 3 m, causing the Black River to dry up. The Cedar River was diverted into the lake through a constructed channel, forming an inlet that now provides about half the inflow to the lake.

Thus, until 1916, smelt from Puget Sound could have had easy access to the lake. Since there is a run of smelt in the Duwamish River now, it seems possible that smelts became isolated in the Cedar River-Lake Washington system in 1916. Even now, the Cedar River population is not conspicuous, so it is conceivable that the fish might have been in the lake between 1916 and 1960, but in such small numbers that it was not noticed. An alternative possibility is that it was introduced in some unknown way shortly before 1960. It would be possible for disoriented smelt to enter the lake from Puget Sound with ships coming through the locks of the ship canal. At present we have no way of deciding which is more likely.

Possible reasons for the increase in smelt

Since predation appears to have been of primary importance in the early links in the sequence under discussion, we looked for data on changes in predators on smelt. Unfortunately there is very little information. The main predator on the smelt and salmon in the 1960s was the northern squawfish (*Ptychocheilus oregonensis*). Data

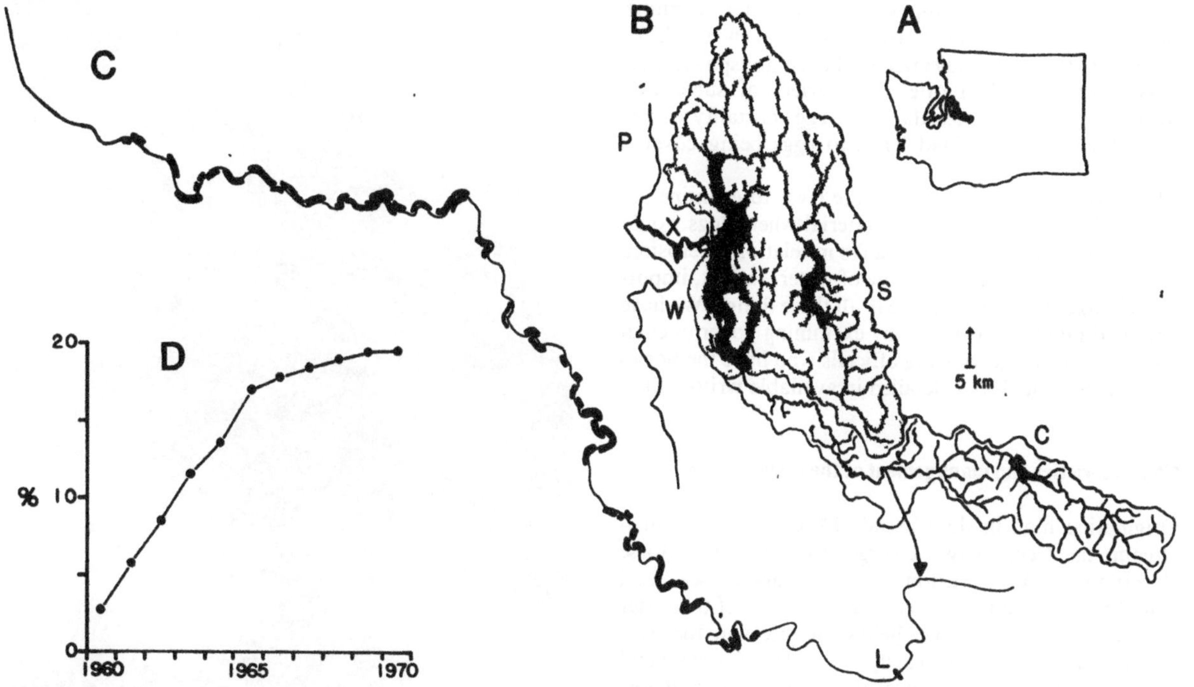

Fig. 1.
A. Map of the State of Washington with Lake Washington watershed shaded.
B. Lake Washington watershed. The unshaded part is the Cedar River watershed. Many small streams are omitted. P, Puget Sound; W, Lake Washington; S, Lake Sammamish; C, Chester Morse Reservoir; L, dam at Landsberg, location of intake for Seattle water supply; X, location of locks in ship canal, the outlet of the lake.
C. Map of lower part of Cedar River. The heavy lines show the location and extent of officially recorded revetments.
D. Graph showing the cumulative percent of the banks of river covered by officially recorded revetments.

on its abundance do not exist, but relative recruitment to Age I class was calculated from the age structure of populations sampled in 1971 and 1972. Recruitment appears to have decreased from the early 1950s to 1963 when it was only about 10% of what it had been in 1957 or 5% of its value in 1952 (BARTOO 1977, see also EGGERS et al. 1978). The reason for the supposed decrease is not obvious. There is no known predator on the squawfish. It is suggestive that the main food species used by the squawfish, the pelagic sculpin (*Cottus aleuticus*), decreased almost to extinction through the 1950s for unknown reasons. Squawfish are opportunistic feeders and have been observed to prey on smelt (OLNEY 1975) and other fish, so it seems unlikely to have decreased for lack of food. Smelt eggs have been found in the gut contents of sculpin in the Cedar River (MOULTON 1970).

Since the record of predation on the smelt is so slight, we considered whether the increase could have been based on increased reproduction or survival of the eggs and young stages in the Cedar River. This idea seems reasonable since the sockeye salmon, which also spawns in the Cedar River, increased in abundance at about the same time as the smelt (EGGERS et al. 1978). While these facts are consistent with an hypothesis based on reproduction and early survival, both species could have increased if both had been released from intense predation by squawfish. Of course, processes affecting both reproduction and mortality could have operated together to permit increased abundance of the two species.

Little or nothing is known about the spawning habits of the longfin smelt, but observations have been made on closely related species (HART & MCHUGH 1944, MCKENZIE 1964). Several investigators note that in spawning areas stream velocities are moderate while the river beds are covered with small gravel or coarse sand. Individual eggs become attached to the gravel by a short stalk and are supported in the water, maximizing oxygen availability. SMITH & SAALFIELD (1955) report that females selectively choose to spawn over gravel bottoms when offered a variety of substrates. MCKENZIE (1964) states that high egg mortality occurs when high sediment load leads to eggs being covered by silt.

The most obvious features of the river to examine are the conditions of the bottom material where eggs would be laid, the condition of the banks in relation to erosion, land use practices, and the flow of water. In fact, important changes have taken place in all these things. We have obtained much information by examining governmental records and by interviewing officials who have memories of activities not clearly detailed in available written files.

Characteristics and management of the Cedar River

Originating high in the Cascade Mountains as a small stream, the Cedar River emerges from Chester Morse Reservoir at an elevation of 510 m above sea level (Fig. 1; see also WILLIAMS et al. 1975). Most of the water comes from melting snow. Below the reservoir the river runs through 24.9 km of heavily forested land past several dams that are used to control the flow of water. Below the dams the river runs in a winding course for 30.1 km of partly agricultural land, and ends with 4.8 km through the small city of Renton at an elevation 7 m above sea level (mean low water). The City of Seattle takes water for domestic use from the Cedar River at Landsburg at the site of the lower dam.

The width of the river bed averages about 20 m. The depth of the river at any point can vary by about 2 m seasonally, and when low can have channel depths of only 0.5 m in places.

During periods of flooding in winter, there was considerable bank erosion, and deposition of silt occurred in the river, creating unfavorable conditions for spawning by fish.

In the 1940s and 1950s the King County Public Works and Commercial Waterway District No. 2 developed a summer program of dredging gravel from the bottom

Fig. 2. An unprotected bank of clay till about 10 m high The water of the river is undercutting the bank, causing large amounts of material to slump into the river from time to time. February, 1987.

Fig 3. The river is protected from an actively eroding bank by a revetment February, 1987.

Fig. 4 A typical revetment along a low bank to prevent private property from washing away during floods. February, 1987.

of the river and piling it along the shore. The purpose was to stabilize the banks against erosion and slumping and to improve the flow of water by deepening the channel. It also removed material that had slumped into the river during the preceding year or had been transported downstream. It seems very likely that this annual disturbance could have damaged the potential spawning beds even though the dredging was not done during spawning season. Conditions for fish were not a consideration at that time since the salmon were very scarce and the smelt unknown.

In 1959, unusually heavy discharge and flooding caused extensive damage to property and slumping of banks. This event motivated King County agencies to stop the dredging program and to institute a program of building revetments at places most susceptible to damage (Fig. 2). The revetments consisted of piles of large stones along the banks (Fig. 3 and 4). Before the protective program started, such structures had already been built at places where a railroad line came near the river, covering nearly 3% of the banks. By the mid 1960s a total of about 20% of the banks in the lower 35 km of the river protected by officially recorded structures (Fig. 1C, D). In addition there are some unrecorded privately built structures that may cover an additional 10%. Thus the increase in salmon and smelt had been accompanied by large changes in conditions known in other cases to affect the success of spawning and survival of small fish including flow of water and condition of banks (Carline & Klosiewski 1985).

Cedar River discharge is dominated by the seasonal cycle of snowfall and melt in the upper watershed, but can be controlled to some extent by the dam at Landsberg. The City of Seattle took all of its drinking water from the Cedar River until 1963 at which time it started using water from the Tolt River also, reducing the amount taken from the Cedar, but not enough to make a pronounced change in discharge. In 1969 the State Legislature passed a law authorizing the Department of Ecology to established standards for minimum flow in the streams to protect fish and other wildlife. Detailed studies have been made of the spawning requirements of salmon in the Cedar River, making it possible to regulate the flow favorably in relation to the life cycle of the salmon (Miller 1976). Although the smelt spawns mostly in February and March, later than the salmon in October to December, it seems likely that it too would benefit from the regulation.

We have less definite information about land use. Presumably in the development of the City of Renton and upstream there was an early period of deforestation, land clearing and property development that would have been completed. There have been various sources of pollution in the lower river, but to some extent they were being ameliorated in the 1960s. For instance, starting in 1942 secondary sewage effluent entered the river a short distance above its entry into the lake, but the effluent was diverted in 1963 (Edmondson & Lehman 1981).

Discussion

At the moment of finishing this paper we have not yet exhausted all the unpublished sources of information about the conditions that permitted the increase of smelt, but we think our study is sufficient to be useful in planning experiments in biomanipulation. If our explanation is correct, an important part of the sequence leading to the outburst of *Daphnia* in Lake Washington was the physical improvement in the Cedar River. This permitted one keystone predator, the smelt, to reduce another keystone predator, *Neomysis*, permitting *Daphnia* to thrive. It is important to distinguish among the various planktonic predators; not all of them will eliminate *Daphnia*.

While each component of the improvement was probably effective, the condition of the gravel where the eggs are laid may be the most important. Because of continuous downstream movement of material, a constant source of gravel is necessary to maintain spawning beds. Too much protection may reduce the spawning area (Reeves & Roelofs 1982). In other situations, perhaps improvement could be made by providing streams with additional areas of suitable substrate and perhaps by making additional spawning channels. Such improvements need not be limited to stream spawners, but can be extended to the littoral region of a lake by suitable procedures.

We have no way of knowing whether the presumed reduction in squawfish was important in permitting the smelt to increase. Therefore we do not know if the improvement in spawning conditions would have been enough by itself to carry out the observed sequence. It seems likely to us that without the improvements in the river, the smelt would not be nearly as successful as it is now.

We do have reason to believe that the decrease in *Neomysis*, although essential, was not enough by itself to permit *Daphnia* to flourish. As long as *Oscillatoria* was abundant, *Daphnia* was scarce (Table 2). Thus, in this particular case, *Daphnia* could not have been used without controlling the nutrient input to the lake. This judgement about the importance of *Neomysis* and *Oscillatoria* is based not simply on correlation, but has a sound basis of experimental work on the effect of the two on *Daphnia*.

Sometimes an environmental manipulation carried out to make a specific improvement in some condition has undesirable side effects. An example is the change in zooplankton following addition of *Mysis relicta*, mentioned in the chapter "Introduction". The present case is different. The revetment program was instituted to protect property against damage by floods, with no consideration of the biological consequences in the Cedar River or in Lake Washington. As it happened, by our interpretation, the changes in the river led to a major improvement in the lake, indicated by increased transparency and improvement in food supply for planktivorous. fish.

Not many lakes will be subject to the same combinations of species and conditions as in Lake Washington, but the major ecological principles will be the same. Our experience suggests that techniques of biomanipulation can be broadened to include alteration of conditions in the watershed designed to improve the success of keystone species in the lake. We hope that this account of the Lake Washington experience will be suggestive and will encourage the use of manipulations as experiments that can help the advance of basic science as well as technology.

Zusammenfassung

Der Lake Washington erlangte 1976 durch eine unerwartete Zunahme an *Daphnia* eine hohe Transparenz. Es zeigte sich, daß sowohl der Predator *Neomysis* als auch die Blaualge *Oscillatoria* (durch Störung des Filtermechanismus) die Daphniendichte gering gehalten hatten. Als die Umleitung von Abwässern die Oscillatoriendichte verringerte und das Auftreten des Fisches *Spirinchus thaleichthyes* zu einer Abnahme von *Neomysis* führte, kam es zur Steigerung der Daphniendichte. Die entscheidende Zunahme der Populationsdichte von *Spirinchus thaleichthyes* wurde offensichtlich durch den höheren Reproduktionserfolg dieser Art ausgelöst Ursache hierfür sind die Verbesserungen der physikalischen Bedingungen im Cedar-River, dem Hauptzufluß zum Lake Washington.

Acknowledgements

The work reported here has been supported by a series of grants from the National Science Foundation and by the ANDREW W. MELLON Foundation.

The following people were very helpful in providing advice or unpublished information in files of governmental agencies: DOUGLAS GROCHOW (King County Surface Water Management), SHELLY MINOR and WILLIAM BOWMAN (King County Public Works), RICHARD HOUGHTON and ROBERT BERGSTROM (Renton Public Works), JOSEPH ROBEL (Washington Department of Fisheries) and ROBERT PFEIFFER (Washington Department of Game). DAVID A. BEAUCHAMP (University of Washington School of Fisheries) provided unpublished information from his studies.

Of the many people who have helped with the Lake Washington study, we make special mention of DAVID ALLISON, KATHLEEN FREVERT, ARNI LITT and MARDI VARELA.

Literature

BARTOO, N. W. (1977): Population parameter estimates and energy budgets for Peamouth, Northern Squawfish, and Yellow Perch in Lake Washington. Ph. D. Diss., Univ. of Washington, 144 p

CARLINE, R. F., & KLOSIEWSKI, S. P. (1985): Responses of fish populations to mitigation structures in two small channelized streams in Ohio. North American Journal of Fisheries Management **5**, 1–11.

DAWIDOWICZ, P., GLIWICZ, Z. M., & GULATI, R. D. (1988): Can *Daphnia* prevent a blue-green algal bloom in hypertrophic lakes? A laboratory test. *Limnologica* (Berlin) **19** (1), 21–26.

DRYFOOS, R. L. (1965): The life history and ecology of the longfin smelt in Lake Washington. Ph. D. Thesis, Univ. of Washington, Seattle, WA, 229 p.

EDMONDSON, W. T. (1980): Secchi disc and chlorophyll Limnol. Oceanogr. **25**, 378–379.

— (1985): Recovery of Lake Washington from eutrophication. Proc. Int. Congress Lakes pollution and recovery (preprints pp. 228–234).

— & LEHMAN, J. T. (1981): The effect of changes in the nutrient income on the condition of Lake Washington. Limnol. Oceanogr. **26**, 1–29.

— & LITT, A. H. (1982): *Daphnia* in Lake Washington Limnol. Oceanogr **27**, 272–293.

EGGERS, D. M, BARTOO, N. W., RICKARD, N A., NELSON, R. E. WISSMAR, R. C., BURGNER, R. L., & DEVOL, A H. (1978) The Lake Washington ecosystem. the perspective from the fish community production and forage base. J. Fish. Res. Bd. Canada **35**, 1553–1571.

EVANS, M. W., & JUDE, D. J. (1986): Recent shifts in *Daphnia* community structure in southeastern Lake Michigan: a comparison of the inshore and offshore regions. Limnol. Oceanogr. **31**, 56–67.

HART, J. L., & MC HUGH, J. L. (1944). The Smelts (Osmeridae) of British Columbia. Fish. Res. Bd. Can. Bull. **64**, 27 pp.

INFANTE, A., & ABELLA, S. E. B. (1985): Inhibition of *Daphnia* by *Oscillatoria* in Lake Washington. Limnol. Oceanogr. **30**, 1046 to 1052.

— & EDMONDSON, W. T. (1985): Edible phytoplankton and herbivorous zooplankton in Lake Washington. Arch. Hydrobiol., Beih. Ergebn. Limnol. **21**, 161–171.

— & LITT, A. H. (1985). Differences between two species of *Daphnia* in the use of ten species of algae in Lake Washington. Limnol. Oceanogr. **30**, 1053–1059.

LAMPERT, W., FLECKNER, W., RAI, H., TAYLOR, B. E. (1986): Phytoplankton control by grazing zooplankton: a study on the spring clear-water phase. Limnol. Oceanogr. **31**, 478–490.

LANGELAND, A. (1981): Decreased zooplankton density in two Norwegian lakes caused by predation of recently introduced *Mysis relicta*. Verh. Int. Verein. Limnol. **21**, 926–937.

MC KENZIE, R. A. (1964): Smelt life history and fishery in the Miramichi River, New Brunswick. Fish. Res. Bd. Can. Bull. **144**, 77 pp.

MILLER, J W (1976): The effects of minimum and peak Cedar River streamflows on fish production and water supply. M. S Thesis, Univ. of Washington, 230 p.

MORGAN, M. D. (1982): Ecology of Mysidacea (Developm. in Hydrobiol. **10**). The Hague—Boston-London [Also: Hydrobiologia **93** (1/2)].

MOULTON, L. L. (1970): The 1970 longfin smelt spawning run in Lake Washington with notes on egg development and changes in the population since 1964. M.S. Thesis, Univ. of Washington, 84 p.

— (1974): Abundance, growth, and spawning of the longfin smelt in Lake Washington. Trans. Am. Fish. Soc. **103**, 46–52.

MURTAUGH, P. A. (1981): Selective predation by *Neomysis mercedis* in Lake Washington. Limnol. Oceanogr. **26**, 445–453.

NIZAN, S, DIMENTMAN, C., & SHILO, M. (1986): Acute toxic effects of the cyanobacterium *Microcystis aeruginosa* on *Daphnia magna*. Limnol. Oceanogr. **31**, 497–502.

OLNEY, F. E (1975): Life history and ecology of the Northern Squawfish *Ptychocheilus oregonensis* Richardson in Lake Washington. M.S. Thesis, Univ. of Washington, Seattle, 75 pp.

PAINE, R. T. (1969): A note on trophic complexity and community stability. Amer. Nat. **103**, 91–93.

— (1980): Food webs: linkage, interaction strength and community infrastructure. J. Animal Ecol. **49**, 667–685.

REEVES, G. H., & ROELOFS, T. D. (1982): Rehabilitating and enhancing stream habitat: 2. Field applications. U.S. Dept. Agriculture, Forest Service, Gen. Tech. Rep. PNW-140, 38 pp.

RIEMAN, B. E. & FALTER, C. M. (1981): Effects of the establishment of *Mysis relicta* on the macrozooplankton of a large lake. Trans. Amer. Fish. Soc. **110**, 613—620.

SCAVIA, D., FAHNSTIEL, G. L., EVANS, M. S., JUDE, D. J., & LEHMAN, J. T. (1986): Influence of salmonine predation and weather on long-term water quality trends in Lake Michigan. Can. J. Fish. Aquatic Sci. **43**, 435—443.

SCHEFFER, V. B., & ROBINSON, R. J. (1939): A limnological study of Lake Washington. Ecol. Monogr. **9**, 95—143.

SCHOENER, T. S. (1983): Field experiments on interspecific competition. Amer. Nat. **122**, 240—285.

SHAPIRO, J. (1979): The need for more biology in lake restoration. In: Lake Restoration. Proceedings of a national conference. Minneapolis. U.S. EPA Rep. No. 440/5-79-001, pp. 161—167.

— (1980): The importance of trophic-level interactions to the abundance and species composition of algae in lakes. In: Hypertrophic Ecosystems. Developm. in Hydrobiol. **2**, pp. 105—116. The Hague.

— & WRIGHT, D. I. (1983): Lake restoration by biomanipulation. Freshwater Biol. **14**, 371—383.

— LAMMARA, V., & LYNCH, M. (1975): Biomanipulation: an ecosystem approach to lake restoration. In: BREZONIK, P. L., & FOX, J. L. (eds.), Proc. symp. on water quality management through biological control. U.S EPA Rep No. ENV-07-75-1, pp. 85—96.

SMITH, W. E., & SAALFIELD, R. W. (1955): Studies on Columbia River smelt (*Thaleichthyes pacificus*). Wash. St. Dept. Fish. Fish. Res. Pap **1** (3), 3—26.

WILLIAMS, R. W., LARAMIE, R. M., & AMES, J. J (1975): Lake Washington Basin. In: A catalogue of Washington streams and salmon utilization. Volume I. Washington State Dept. of Fisheries.

WRIGHT, D. I., & SHAPIRO, J., (1984): Nutrient reduction by biomanipulation: an unexpected phenomenon and its possible cause. Verh. Int. Verein. Limnol. **22**, 518—524.

WYDOSKI, R., & WHITNEY, R. R. (1979): Inland fishes of Washington. Seattle, London, 220 p.

Received March 27, 1987

Address of the authors:

Prof. W. T. EDMONDSON
SALLY E. B. ABELLA
Department of Zoology, NJ-15
University of Washington
Seattle, WA 98195
U.S.A.

Limnologica (Berlin) 19 (1), 81—85, Juli 1988

Limnological Institute, Oosterzee, The Netherlands

Trophic Interactions in the Hypertrophic Lake Tjeukemeer: Top-down and Bottom-up Effects in Relation to Hydrology, Predation and Bioturbation During the Period 1974—1985

By Eddy H. R. R. Lammens

With 3 Figures and 2 Tables

Key words: Chlorophyll-a; *Daphnia hyalina*, Predation; Bream; Hydrology

Abstract

During the summer months of 1974—1985 chlorophyll-a and total P concentration, average size and density of *Daphnia hyalina*, biomass of smelt *Osmerus eperlanus* and bream *Abramis brama* and water temperature were monitored. A multivariate-regression-analysis showed a top-down effect between smelt and *D. hyalina*, in particular regarding the average size, and between bream and chlorophyll-a, and a bottom-up effect between total P and chlorophyll-a and between chlorophyll-a and *D. hyalina* Significant negative correlations between chlorophyll-a on one side and water temperature and smelt on the other were probably related to the hydrology of the lake, which is largely determined by temperature and precipitation During summer relatively nutrient poor IJsselmeer water enters the lake and dilutes the nutrient rich water, which has entered the lake in winter from the surrounding fertilized agricultural areas.

Introduction

One of the central issues in limnology is eutrophication. Most lakes and rivers in the temperate zone experience this process, due to an overload of nutrients from several sources. In particular in very densely populated countries such as the Netherlands lakes have lost their original status of clear vegetation-rich waterbodies with a diverse fish community. Instead the systems have become turbid, dominated by blue-green algae (Moed & Hoogveld 1982) and a few fish species dominate, such as bream and pike-perch (Vijverberg & Van Densen 1984, Lammens et al. 1985, Lammens 1986). At the moment there are two approaches to reverse the eutrophication process, that is, reducing the input of nutrients and enlarging the predation pressure on phytoplankton by manipulation of the food web (see for a review McQueen et al. 1986).

During this twelve year study (1974—1985) the fishery management of Tjeukemeer has been changed by the ending of the gill-net fishery for pike-perch in 1977. This fishery removed about 70% of the pike-perch >42 cm (Goldspink & Banks 1975) and a not known part of the bream population >28 cm yearly. Since this change in 1977 the pike-perch and bream populations have changed significantly with respect to the larger length-classes and the larger zooplankton species have changed as well (Lammens 1986). During the same time, however, the hydrological management has changed too, as, apart from temperature effects, decreasing amounts of relatively nutrient-poor IJsselmeer water were let in. In order to explain the development of the phytoplankton (as a function of chlorophyll-a) and the zooplankton (*D. hyalina*) during this period multivariate analyses were performed with the summer values of chlorophyll-a, and average size and density of *Daphnia hyalina* as dependent variables. As independent variables were chosen the biomass of the bream and smelt population, temperature, yearly water intake from IJsselmeer, total P and year.

Study Area

Tjeukemeer is a shallow (depth 1.5—2 m), hypertrophic and turbid (Secchi-disc 25—35 cm) freshwater lake with a surface area of 21.5 km^2. The total P concentration amounts to 0.25 mg/l in summer and to 0.3 mg/l in winter, whereas total N amounts to 3 and 6 mg/l, respectively. The inshore zone is poorly developed, the index of shore-line development being 1.25 (Hutchinson 1957). The only vegetation is some reed. Most of the primary production is realized by the blue-green alga *Oscillatoria agardhii* (Moed & Hoogveld 1982) and most of the secondary production by cladocerans and copepods, of which *Daphnia hyalina* is most productive (Vijverberg & Richter 1982a, b). The production and biomass of chironomids is very small (Beattie 1982). The fish community is dominated by bream (*Abramis brama*) and pike-perch (*Stizostedion lucioperca*), whereas smelt (*Osmerus eperlanus*) is present in varying amounts (Lammens 1986). As the lake is shallow and exposed to wind, summer stratification does not occur and as a result zooplankton is almost homogeneously distributed (Nie et al. 1978, Nie & Vijverberg 1985). The lake is part of an interconnected system of lakes, that receives water from the nearby lake IJsselmeer during the summer and from the surrounding polders in winter, when precipitation exceeds evaporation (Fig. 1). The management of the hydrological regime is designed to control water level which is required for agricultural purposes. Chemistry and hydrology have been described by De Haan (1982), De Haan & Moed (1984) and Leenen (1982).

Material and Methods

Fish and zooplankton were collected during a period of 12 years (1974—1985). The fish were caught once a month in summer with a small mesh trawl (5.5 mm cod end) a five stations. Standard hauls were made with a speed of 1 m/sec during 10 min. Length-frequency distribution and total number of each species were determined. Yearly catches of bream are given as an average for the whole summer, that is 4 times 5 seperate catches. However, because smelt (*Osmerus eperlanus*) is not caught efficiently with this trawl until the second half of the summer, only the September catches of this species were used.

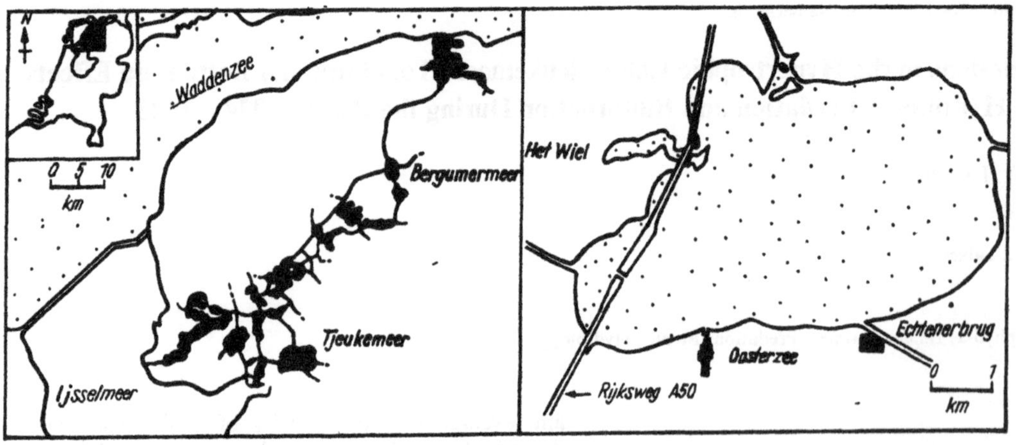

Fig. 1. The location of Tjeukemeer in the Friesian lake district and The Netherlands. Lake IJsselmeer is shown as well.

Zooplankton was sampled at the same time and place of trawling. Using 1 5-l Friedinger sampler with a cylinder of 60 cm height two samples were taken at each of the five stations, one just below the water surface and the other just above the bottom. The samples were concentrated by filtering through a 120 µm mesh plankton gauze and preserved in 4% formalin. The combined sample was sub-sampled with a whirling vessel after KOTT (1953) and organisms were identified, counted and their lengths measured. The length of the cladocerans was measured from the top of the head or, if a helmet was present, from the base of the helmet to the base of the tail spine

Chlorophyll-a and total P concentrations were determined fortnightly in a mixed sample of ten stations. Chlorophyll-a was determined according to MOED (1973) and the concentration was calculated after NUSCH & PALME (1975). Total P was analysed after hydrolysis to ortho-phosphate for 2.5 h at 110 °C in acidic (3N H_2SO_4) persulphate (5%). Ortho-phosphate was analysed colorimetrically (MURPHY & RILEY 1962) either manually or using an auto-analyser (Technicon AA II). Water temperature was measured weekly from three whole-column samples. The inflow of IJsselmeer water was recorded by the department of Public Works in the province of Friesland.

Results and Discussion

During the period 1974—1985 the chlorophyll-a concentration in the lake increased significantly ($p < 0.05$), together with the average size and density of *D. hyalina* ($p < 0.01$ and $p < 0.05$ resp.) and the biomass of the bream population ($p < 0.001$), whereas the water intake from lake IJsselmeer decreased during this period ($p < 0.005$) (Table 1, Fig. 2). Temperature, total P and smelt did not show a significant trend during this time period.

Daphnia hyalina as dependent variable

The average size and density of *D. hyalina* showed positive correlations ($p < 0.005$ resp. $p < 0.01$) with chlorophyll-a and weaker negative correlations with smelt biomass ($p < 0.05$ resp. $p > 0.1$). In a multiple regression analysis smelt predation and chlorophyll-a explained about 51% of the total variation of the average length of *D. hyalina*, which could not be increased by adding the other variables (Table 2, Fig. 3). About 36% of the variation in density was explained by chlorophyll-a, which could not be increased by adding other variables. The positive effect may be caused by a direct (bottom-up trophic interaction) or indirect (some unkown common factor) relationship (Table 2, Fig. 3). Because only the September smelt biomass is used, its impact on *D. hyalina* is biased and its contribution in the variation of size and density of *D. hyalina* may be greater.

The trophic interaction between chlorophyll-a and *D. hyalina* is complex as the latter shows a negative selection for *Oscillatoria* (BLOEM & VIJVERBERG 1984) because it has problems handling these organisms (GULATI unpublished, see also DAWIDOWICZ et al. 1988). Due to this negative selection an increasing predation pressure of *D. hyalina* may indirectly stimulate *Oscillatoria* as other organisms are selectively removed, giving more 'room' to *Oscillatoria*. The resulting positive correlation may be similar to that of the predatory fish pike-perch and its least preferred prey-fish bream, as the latter meets a better feeding condition when preferred prey-fish are removed by pike-perch (LAMMENS 1986). A negative correlation between *D. hyalina* and chlorophyll-a is only probable when all phytoplankton

Table 1. Correlation matrix of of variables.

	Average size	Density *D. hyalina*	Biomass smelt	Biomass bream	Water Temper.	Year	Water intake	Total P
Chlorophyll-a	0.732**	0.647*	−0.601*	0.607*	−0.172	0.638*	−0.638*	0.642*
Av. size *D. hyalina*		0.813***	−0.644*	0.427	−0.069	0.609*	−0.522	0.135
Density *D. hyalina*			−0.460	0.381	−0.160	0.679*	−0.572*	0.207
Biomass smelt				0.207	−0.396	−0.383	0.090	−0.191
Biomass bream					0.062	0.824****	−0.574*	0.613*
Temperature						0.009	0.555*	0.090
Water intake from IJsselmeer							−0.746***	0.447
Total P								−0.211

* = $p < 0.05$, ** = $p < 0.01$, *** = $p < 0.005$, **** = $p < 0.001$

organisms can easily be eaten. PORTER (1977) showed that several feed-back mechanisms occur in phytoplankton-zooplankton interactions as a result of feeding selectivity and are essential in the structure of these communities. This study shows clearly that phytoplankton is not limited by the *D. hyalina* population, but rather the other way around.

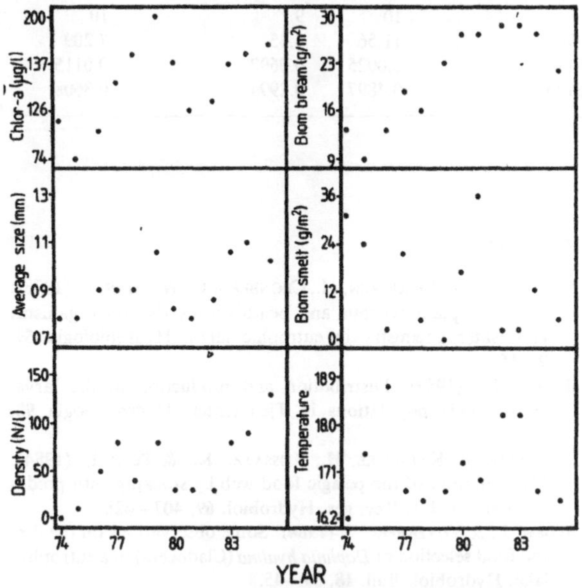

Fig 2. Density and average size of *Daphnia hyalina*, chlorophyll-a concentration, biomass of bream and smelt, and temperature are plotted against the year. For five variables average summer values are given, for smelt only September values are given

Chlorophyll-a as dependent variable

Temperature and smelt biomass showed weak negative correlations with chlorophyll-a (Table 1), but in a multiple regression analysis these (partial) correlations became very significant (Table 2). About 84% of the variation of chlorophyll-a could be explained by temperature (negative), total P (positive), bream (positive) and smelt (negative). The negative relation to temperature and smelt is not expected as both variables are supposed to have the opposite effect. To understand the negative relation to temperature and smelt one has to know more about the hydrology of the Friesian lake district and about the limnology of the surrounding waters. The Friesian lake district (Fig. 1) is an interconnected system of lakes, in which the water level is regulated for agricultural purposes. In summer there is usually a water deficiency because evaporation is greater than precipitation. Then water enters the lakes from lake IJsselmeer (Fig. 1). In winter there is mostly too little evaporation and water is drained from the fertilized agricultural areas and pumped into the lakes and from the lakes into the IJsselmeer and the Waddensea. However, as temperature and precipitation differ between years this has consequences for the input of nutrients and smelt. In relation to the polder waters the nutrient levels in the IJsselmeerwater are low and cause much lower chlorophyll-a levels than the enriched polder waters (DE HAAN & MOED 1984).

The negative interaction between chlorophyll-a and smelt is in contrast to what is expected in a top-down interaction. The large impact of the hydrology on the abundance of smelt larvae overrules this top-down effect

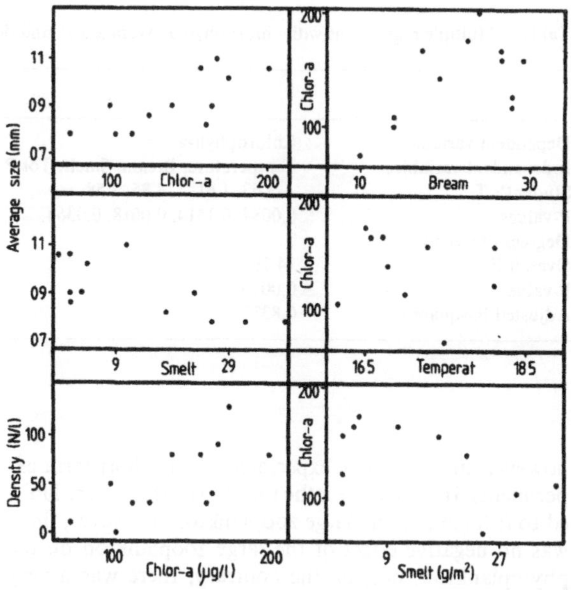

Fig. 3. Average summer values of chlorophyll-a concentration are plotted to biomass of bream, smelt and temperature, and average size and density of *Daphnia hyalina* are plotted to chlorophyll-a concentration and smelt biomass.

completely: smelt larvae are largely born in the lake IJsselmeer and passively transported to the lake district (VAN DENSEN & VIJVERBERG 1982). The relatively nutrient-poor IJsselmeer water is 'smelt'-rich and probably therefore chlorophyll-a and smelt show a negative relation.

During these fourteen years the water intake from lake IJsselmeer has significantly decreased, whereas the biomass of bream has increased (Table 1); both processes have a stimulating effect on the availability of nutrients. The ending of the pike-perch gill-net fishery at the beginning of 1977 caused this significant increase of bream > 30 cm (LAMMENS 1986), whereas the decreasing water intake was probably related to the absence of botulism in the eighties, which required extra water intake in the seventies.

The current hypothesis about top-down effects

Although there is a marked influence of the predation pressure of smelt on *D. hyalina*, this effect does not mask the influence of the nutrients. The negative effect of smelt on chlorophyll-a concentration is clearly not a direct (top-down) effect, otherwise the relation should have been positive, but an indirect relation to the hydrology as is the effect of temperature. These results are somewhat contradictory with findings in literature concerning the role of fish and zooplankton in the dynamics of phytoplankton, that is the top-down effect (ANDERSSON et al. 1978, BENNDORF et al. 1984, SHAPIRO & WRIGHT 1986, BRABRAND et al. 1986, MCQUEEN et al. 1986). The general idea is that a strong predation pressure on the zooplankton changes the zooplankton community into small specimens and species and thus stimulates eutrophication because of the relieved predation pressure on the phytoplankton. This is often found in enclosure experiments and fish ponds, which were manipulated to show these interactions. In many of these experiments top-down effects up to the nutrient level were obvious,

Table 2. Multiple regressions with chlorophyll-a, average size and density of *Daphnia hyalina* as dependent variables.

	1	2	3	4	5
Dependent variable	Chlorophyll-a	Size *D hyalina*	Idem	Density *D. hyal.*	Idem
Independent variables	Temperature, Bream, Smelt, Tot P	Chl.-a, Smelt	Chl.-a	Chl.-a, Smelt	Chl.-a
Student's T	—3.63, 1.61, —4.85, 2.58	2.05, —1.21	3.40	1.84, —0.28	2.68
P-values	0.0084, 0.1514, 0.0018, 0.0364	0.07, 0.257	0.0068	0.099, 0.733	0.0229
Degrees of freedom	7	9	10	9	10
Overall F	14.79	6.778	11.56	3.351	7.209
P-value	0.0013	0.0110	0.0025	0.0692	0.0115
Adjusted R-squared	0.8337	0.5123	0.4897	0.2994	0.3608

however, most of these experiments were short term experiments. In Tjeukemeer the top-down effects were limited to fish (smelt) and large zooplankton. However, there was no negative effect of the large zooplankton on the phytoplankton, but, on the contrary, there was a very significant positive relation between size and density of *D. hyalina* and chlorophyll-a and therefore a bottom-up effect. So instead of being positively related to smelt biomass and temperature the opposite was true and only bream showed a significant positive correlation with phytoplankton and total P. The benthic foraging of this fish may have positively affected the release nutrients from the bottom sediments and thus stimulated phytoplankton growth, as is shown in some experiments (Andersson et al. 1978, Brabrand et al. 1986, Tatrai & Istvanovics 1986). On top of this the simultaneous decrease of water intake from lake IJsselmeer may have the same effect, due to its diluting effect. This was, however, not clear from the statistical analysis, whereas bream explained a larger part of the variation in chlorophyll-a than water intake from IJsselmeer did.

It is clear that both bottom-up and top-down effects are present in Tjeukemeer, but the bottom-up effects are more important than the top-down effects, due to the large impact of hydrology on this lake. Because of this dominating influence it is very difficult to reverse the eutrophication process by manipulating the food web. Moreover, in a recent review article McQueen et al. (1986) showed that top-down effects are only obvious in less eutrophic lakes, whereas these effects are only very limited in highly eutrophic lakes. Removing the bream population would probably show some effect, however, this does not seem realistic.

Zusammenfassung

Für die Jahre 1974—1985 (Sommermonate) werden die ermittelten Beziehungen zwischen dem Chlorophyll-a-Gehalt, der durchschnittlichen Körpergröße und der Populationsdichte von *Daphnia hyalina*, der Biomasse der Fische *Osmerus eperlanus* und *Abramis brama* sowie der Wassertemperatur in Form einer multivarianten Regressionsanalyse dargestellt (Fig. 3).

Acknowledgements

I thank Henk De Haan and Jan Moed for some of their unpublished results of chlorophyll-a concentrations, total P and temperatures and for their valuable comments on the manuscript. Aafje Frank-Landman, Bertus Lemsma, and Koos Swart are acknowledged for assistence in the field and Jürgen Benndorf, Wim Van Densen, Louis Van Liere, Sikko Parma, Bert Richter, Adee Schoon and Koos Vijverberg for critical comments.

References

Andersson, G., Berggren, H., Cronberg, G., & Gelin, C. (1978): Effect of planktivorous and benthivorous fish on organisms and water chemistry in eutrophic lakes. Hydrobiologia **59**, 9—15.

Beattie, M. (1982): Distribution and production of the larval chironomids populations in Tjeukemeer Hydrobiologia **95**, 287—306

Benndorf, J., Kneschke, H., Kossatz, K., & Penz, E. (1984): Manipulation of the pelagic food web by stocking with predacious fishes. Int. Rev. ges. Hydrobiol. **69**, 407—428.

Bloem, J., & Vijverberg, J. (1984): Some observations on the diet and food selection of *Daphnia hyalina* (Cladocera) in a eutrophic lake. Hydrobiol. Bull. **18**, 39—45.

Brabrand, A., Faafeng, B., Nilssen, J. P. M. (1986): Juvenile fish and invertebrate predators: delaying the recovery phase of eutrophic lakes by suppression of efficient filter feeders. J. Fish Biol. (in press).

Dawidowicz, P., Gliwicz, Z. M., & Gulati, R. D. (1988). Can *Daphnia* prevent a blue-green algal bloom in hypertrophic lakes? A laboratory test. *Limnologica* (Berlin) **19** (1), 21—26.

De Haan, H. (1982): Physico-chemical environments in Tjeukemeer with special reference to speciation of algal nutrients. Hydrobiologia **95**, 205—221.

— & Moed, J. R. (1984): Phosphorus, nitrogen and chlorophyll-a concentrations in a typical Dutch polder lake, Tjeukemeer in relation to its water regime between 1968 and 1982. Wat. Sci. Technol. **17**, 733—743.

Densen, W. L. T. van, & Vijverberg, J., (1982): The relations between O+fish density, zooplankton size and vulnerability of pike-perch, *Stizostedion lucioperca*, to angling in the Friesian lakes. Hydrobiologia **95**, 321—336.

Goldspink, C. R., & Banks, J. W. (1975): A description of the Tjeukemeer fishery together with a note upon the yield statistics between 1964 and 1970. J. Fish Biol. **7**, 687—708.

Hutchinson, G. E. (1957): A treatise on limnology II New York.

Kott, P., 1953. A modified whirling apparatus for subsampling of plankton. Aust. J. Mar. Freshwat. Res. **4**, 387—393.

Lammens, E. H. R. R. (1986): Interactions between fishes and the structure of fish communities in Dutch shallow eutrophic lakes. Ph. D. Thesis, 100 p.

— De Nie, H. W., Vijverberg, J., & Densen, W. L. T. van (1985): Resource partitioning and niche shifts of bream (*Abramis brama*) and eel (*Anguilla anguilla*) mediated by predation of smelt (*Osmerus eperlanus*) on *Daphnia hyalina*. Can. J. Fish. Aquat. Sci. **42**, 1342—1351.

Leenen, J. D. (1982): Hydrology of Tjeukemeer. Hydrobiologia **95**, 199—203.

Lynch, M. (1979): Predation, competition and zooplankton community structure: an experimental study. Limnol. Oceanogr. **24**, 253—272.

McQueen, D., Post, J. R., & Mills, E. L. (1986): Trophic relationships in pelagic ecosystems. Can J. Fish. Aquat. Sci. **43**, 1571—1581.

Moed, J. R. (1973): Effect of the combined action of light and silicon depletion on *Asterionella formosa* Hass. Verh. Int. Verein. Limn. **18**, 1367—1374.

— & HOOGVELD, H. L. (1982): The algal periodicity in Tjeukemeer during 1968—1978. Hydrobiologia **95**, 205—223.

MURPHY, J., & RILEY, J. P. (1962): A modified single-solution method for the determination of phosphate in natural waters. Anal. Chim. Acta **27**, 31.

NIE, H. W. DE, & VIJVERBERG, J. (1985): The accuracy of population density estimates of copepods and cladocerans, using data from Tjeukemeer (The Netherlands) as an example. Hydrobiologia **124**, 3—11.

— BROMLEY, H J., & VIJVERBERG, J. (1978): Distribution patterns of zooplankton in Tjeukemeer, The Netherlands. J Plankt Res **2** (4), 317—334.

NUSCH, E., & PALME, G. (1975): Biologische Methoden für die Praxis der Gewässeruntersuchung. GWF-Wasser/Abwasser **116**, 562—565.

PORTER, K. G. (1977): The plant-animal interface in freshwater ecosystems. Am. Sci. **65**, 159—170.

SHAPIRO, J., & WRIGHT, D. I. (1986): Lake restoration by biomanipulation: Round Lake, Minnesota, the first two years. Freshwat. Biol. **14**, 371—383.

TATRAI, I., ISTVANOVICS, V. (1986): The role of fish in the regulation of nutrient cycling in lake Balaton Hungary Freshwat. Biol. **16**, 417—424.

VIJVERBERG, J, & RICHTER, A. F. (1982a): Population dynamics and production of *Daphnia hyalina* LEYDIG and *Daphnia cucullata* SARS in Tjeukemeer. Hydrobiologia **95**, 235—259.

— — (1982): Population dynamics and production of *Acantocyclops robustus* SARS and *Mesocyclops leucarti* CLAUS Hydrobiologia **95**, 261—274

— & DENSEN, W. L. T. VAN (1984): The role of the fish in the food web of Tjeukemeer. Verh Int Ver Limnol. **22**, 891—896.

Received January 5, 1987

Address of the author:

Dr. EDDY H. R. R. LAMMENS
Limnological Institute
de Akkers 47
8536 VD Oosterzee
The Netherlands

Limnologica (Berlin) 19 (1), 87—95, Juli 1988

AdW der DDR, Zentralinstitut für Mikrobiologie und Experimentelle Therapie, Abteilung Limnologie, Jena, Technische Universitat Dresden. Sektion Wasserwesen, Bereich Hydrobiologie, Dresden, DDR

Applicability of the Food-web Manipulation in the Restoration Program of a Hypertrophic Stratified Lake: Model Studies for Lake Haussee (Feldberg, GDR)

By PETER KASPRZAK, JÜRGEN BENNDORF, RAINER KOSCHEL & FRIEDER RECKNAGEL

With 6 Figures and 2 Tables

Key words: Food-web manipulation; Sewage diversion; Calcite precipitation; Scenario-Analysis

Abstract

Lake Haussee (Feldberg, GDR) is a highly eutrophic lake due to external nutrient load from a municipial waste-water treatment In 1980 the effluent was interrupted, but the lake remained in the state of hypertrophy because of the great amount of accumulated nutrients and high internal load. Investigations about hydrology and chemistry on the one side and on the structure of the food-web on the other side showed, that the self-purification potential is rather low. Therefore we investigated the effect of food-web manipulation by means of scenario studies using the ecological lake model SALMO.

It was shown that decreased zooplankton mortality leads to drastic reduction of phytoplankton biomass. The second part of the spring maximum is nearly suppressed and also the concentration throughout the rest of the year is 50—75% lower. There are indications for interactions between food-web manipulation and calcite precipitation, another important loss-factor of phytoplankton in hard-water lakes. Pelagic calcite precipitation with its connected decrease of primary production works most efficient in mesotrophic lakes. Because food-web manipulation leads also to limitation of phytoplankton growth a stimulation of calcite precipitation can be assumed.

Introduction

It has been well known for some time that the trophic state of waters depends not only on the nutrient load but also on the structure of the food chain. The amount and the quality of phytoplankton, e.g., is strongly influenced by zooplankton biomass and that by the predation pressure of planktivorous fish (UHLMANN 1954, HRBÁČEK 1958, HRBÁČEK et al. 1961). The idea to use these interactions for water quality management has been stressed more and more since the midseventies (e.g. SHAPIRO et al. 1975, STEEL 1975, ANDERSSON et al. 1978, STENSON et al. 1978, BARTHELMES 1978, BENNDORF & RECKNAGEL 1978, BENNDORF 1979). The most important results can be summarized as follows:

(1) Promoting the stock of predacious fish it is possible to lower the biomass of planktivorous fish (HRBÁČEK et al. 1977, BENNDORF et al. 1984, 1985, 1988, HESSEN 1983). Zooplankton-feeding fish are able to decimate the zooplankton drastically. In a mixture of different sized plankters they select the biggest. Therefore the decrease of their biomass leads to the growth of big filtrators (BROOKS & DODSON 1965, O'BRION 1979, ZARET 1980, BENNDORF et al. 1984).

(2) The role of invertebrate predators must be differentiated. Predacious copepods and *Chaoborus* prefer small and mediumsized prey (BRANDL & FERNANDO 1975, KERFOOT 1977, FEDORENKO 1975, KAJAK & RYBAK 1979) and promote therefore the desired effect. The consequences of high abundances of *Leptodora* and *Neomysis* can be quite negative because these species have no clear preference for small prey (KARABIN 1974, GLIWICZ et al. 1978, MURTAUGH 1981). For this reason food-web manipulation does not have the aim of exterminating the planktivorous fish population because the control of the big invertebrate predators must be guaranteed (STENSON 1978, BENNDORF et al. 1984, 1987, EDMONDSON & ABELLA 1987).

(3) Also in large reservoirs and lakes herbivorous zooplankton has a strong influence on the biomass of phytoplankton (DUNCAN 1975, LAMPERT & SCHOBER 1978, HORN 1981, KASPRZAK & RONNEBERGER 1985).

(4) BRIAND & MCCAULEY (1978) have concluded theoretically and later shown in experiments (MCCAULEY & BRIAND 1979) that the structure of phytoplankton community changes to big uningestable forms because filtrators prefer to feed on small algae. Moreover there are small phytoplankters whose growth rates are high enough to resist the increased grazing pressure (FOTT 1975, BENNDORF et al. 1984).

Meanwhile there are a number of investigations indicating the usefulness of the concept of food-web manipulation as a measure for water quality management (SHAPIRO et al. 1975, HRBÁČEK et al. 1977, STENSON 1978, BJÖRK et al. 1978, ANDERSSON et al. 1978, FOTT et al. 1980, LEAH et al. 1980, HENRIKSON et al. 1980, EDMONDSON & LITT 1982, BENNDORF et al. 1984, GOAD 1984, SHAPIRO & WRIGHT 1984, REINERTSEN & OLSEN 1984). The uncertainty concerns the scale of the experiments which extends from big plastic tubes up to whole lakes. Origin, amount and kind of nutrient load have been very different. Frequently the effect of food-web manipulation was unintended or the whole fish stock was simply poisoned. Introduction of predacious fish was an exception. Therefore it is uncertain if the improvements of water quality published in serveral experiments (decrease of phytoplankton biomass, increased Secchi depth, reduced pH, decrease of total phosphorus, dominance of big herbivorous zooplankters) will also occur in waters with a high phosphorus load.

Current investigations have shown that in the case of high external phosphorus load not all the desired improvements were achieved (BENNDORF et al. 1988). The same seems to be true for shallow unstratified waters with high internal phosphorus load (LAMMENS 1988).

However, there exists no experience of the possibilities of food-web manipulation in highly eutrophic stratified hard-water lakes with low external but high internal phosphorus load from the sediment and low renewal rate of the water. But one can assume that the influence of increased filtration performance of zooplankton has other results compared with highly flushed reservoirs or lakes pure in calcium carbonate. In highly eutrophic hard-water lakes there is the possibility of intensified autochthonous calcite precipitation due to food-web manipulation with its connected reactions leading to an improvement in water quality (Koschel et al. 1983, 1987).

For these reasons we selected the $CaCO_3$-rich, highly eutrophic Lake Haussee for such an experiment. The aim is to force this lake to a lower level of trophy by means of introducing pike-perch (*Stizostedion lucioperca* L., 200 one-summer-old fishes per ha) and selective fishing of planktivorous species.

However, to begin with we want to investigate the effects of food-web manipulation on the trophic character of the lake theoretically by means of scenario studies using the ecological lake model SALMO (Benndorf 1979, Recknagel 1980). The model was tested with success in several practical problems of decision-making in eutrophication management (cf. Benndorf & Recknagel 1982, Benndorf et al. 1985, Recknagel 1985, Benndorf 1985).

For our problem SALMO can give results in sense of relative comparisons of state variables (phosphate, inorganic nitrogen, oxygen, two functional groups of phytoplankton, herbivorous zooplankton, detritus). Because of the present relatively simple structure of SALMO it is impossible to make studies of structuring effects of food-web manipulation.

All the methods applied to the measurement of data quoted in the paper are described in Koschel et al. (1981).

Table 1. Lake Haussee, hydrographical, hydrochemical and hydrobiological features; 1—9 according to Koschel et al. (1985), 10, 11 according to Koschel et al. (1987), except Secchi depth and phytoplankton.

1	area (km^2)	1.36
2	maximum depth (m)	12.00
3	mean depth (m)	6.00
4	volume ($m^3 \cdot 10^6$)	8.15
5	catchment area (km^2)	4.00
6	theoretical water renewal time (a)	5.00
7	nutrient load	
	before waste-water diversion	
	P ($gP \cdot m^{-2} \cdot a^{-1}$)	1.6— 2.1
	N ($gN \cdot m^{-2} \cdot a^{-1}$)	8.9—13.4
	after waste-water diversion 1980	
	P ($gP \cdot m^{-2} \cdot a^{-1}$)	0.13—0.28
	N ($gN \cdot m^{-2} \cdot a^{-1}$)	1.5—3.1
8	nutrient concentration at the beginning of spring circulation mean values 1978—1982	
	PO_4-P ($gP \cdot m^{-3}$)	1.1
	TP ($gP \cdot m^{-3}$)	1.3
	NO_3-N ($gN \cdot m^{-3}$)	0.5
	NH_4-N ($gN \cdot m^{-3}$)	0.1
9	planktonic primary production ($gC \cdot m^{-2} \cdot a^{-1}$)	410
10	use of primary product by planktonic crustaceans (%)	8—9
11	annual mean values of euphotic zone 1978—1982	
	Secchi depth (m)	1.5
	phytoplankton ($cm^{-3} \cdot m^{-3}$)	7.3*
	pH	8.9
	Ca ($g \cdot m^{-3}$)	71.0
	$CaCO_3$ ($mg \cdot m^{-3}$)	80.0
	DIC ($gC \cdot m^{-3}$)	36.0

*) conversion factor from chl-a = 500

Limnological state analysis of Lake Haussee

Table 1 shows some important criteria of the hydrography and the water quality of Lake Haussee. At the beginning of our century the lake was slightly eutrophic having a mean summer Secchi depth of about 5 m and also in August an aerobic hypolimnion (Thienemann 1925). But already the results of Ohle (1934) from August 1932 showed a clear deterioration due to anthropogenic nutrient load. This tendency was observable also in the following decades (Koschel et al. 1981, 1985) and reached its maximum at the beginning of the sixties as the town of Feldberg acquired a municipial waste-water treatment. 90% of the external nutrient load of the lake originated now from the effluent (Hamann cit. a. Koschel et al. 1981), and Lake Haussee was changed into a highly eutrophic lake.

The first step in the restoration program was the diversion of the treatment effluent in 1980. In this way the nutrient load was decreased drastically and is now within a range which can be tolerated for stratified lakes of moderate eutrophy (cf. Koschel et al. 1985).

However the nutrients enrichment over decades acts against fast improvement of the water quality. A comparison of some parameters from 1978 until 1985 shows that Lake Haussee is still a nutrient rich, highly eutrophic lake (Fig. 1). The spring maximum of dissolved inorganic phosphorus (DIP) shows only a slight but significant decrease (T-test, $P = 0.05$). The rate of decrease amounts to $70\ mgP \cdot m^{-3} \cdot a^{-1}$. Phosphate limitation of primary production can be excluded because of the present concentration of about $800\ mgP \cdot m^{-3}$. Much smaller is the decrease of total phosphate (TP). It amounts only to $30\ mgP \cdot m^{-3} \cdot a^{-1}$ and is statistically insignificant. Using a model of a well mixed reactor and the rate just cited, it would take about 40 years until Lake Haussee reached a level of moderate eutrophy with a spring maximum of TP of about $200\ mgP \cdot m^{-3}$ (cf. Koschel et al. 1981, 1985). It is difficult to explain the differences in the decreasing rate of both phosphorus fractions. The slow diminution of TP indicates that the decrease of DIP is caused only to a small extent by dilution processes following the renewal of water. We assume that the reason is the different start of spring bloom of phytoplankton and the connected consumption of DIP. Evidence for the truth of our assumption comes from equal DIP and TP concentrations in 1978, 1981 and 1984. In those years the samples have been taken during low phytoplankton concentrations.

The annual mean concentration of phytoplankton shows great scatter (cf. Fig. 1). A decreasing tendency is not detectable. Until 1980 chlorophyceans prevailed in the plankton but were now increasingly replaced by blue-greens. The annual mean Secchi depth is unchanged lower than two meters, the pH is around 8.9, the Ca-content amounts to $71\ g \cdot m^{-3}$ and the concentration of inorganic dissolved carbon to $36\ g \cdot m^{-3}$. Therefore we can conclude that in the coming years a rapid improvement of trophic situation in Lake Haussee cannot be expected because of the high nutrient concentrations. The regeneration process can be accelerated by anti-eutrophication measures, in our case by means of food-web manipulation.

Figs. 2 and 3 show the annual distribution of crustacean biomass and the corresponding mean body weights. Both parameters are of special interest for food-web manipulation because they express the current situation

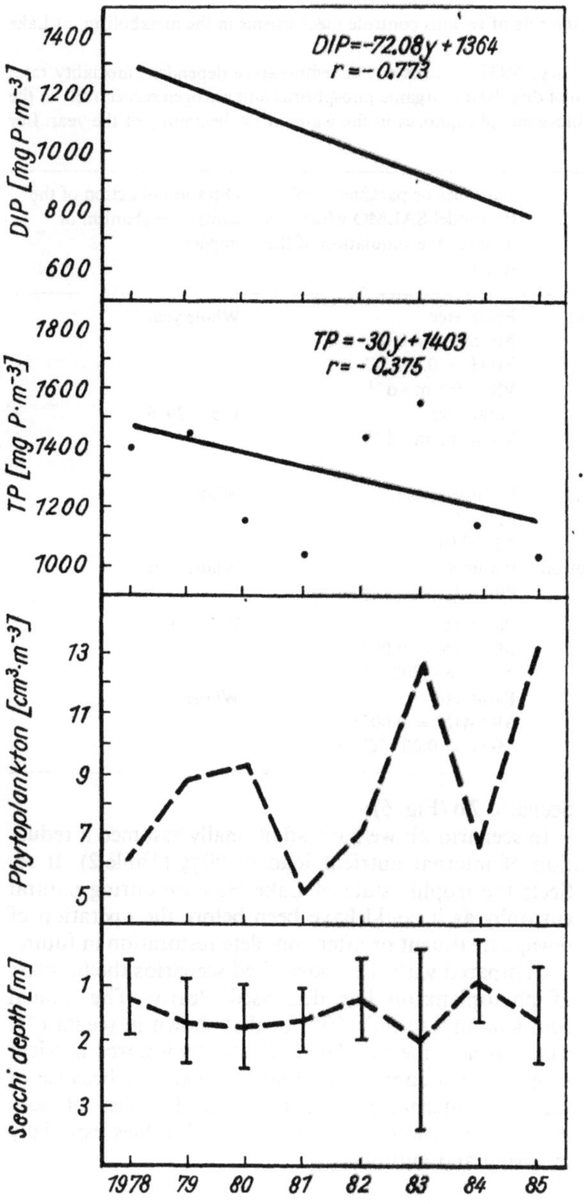

Fig 1. Selected features of water quality in Lake Haussee.

TP — total phosphorus, DIP — dissolved inorganic phosphorus Phytoplankton concentration is based on Chl-a values using a conversion factor of 500, Secchi depth with standard deviation

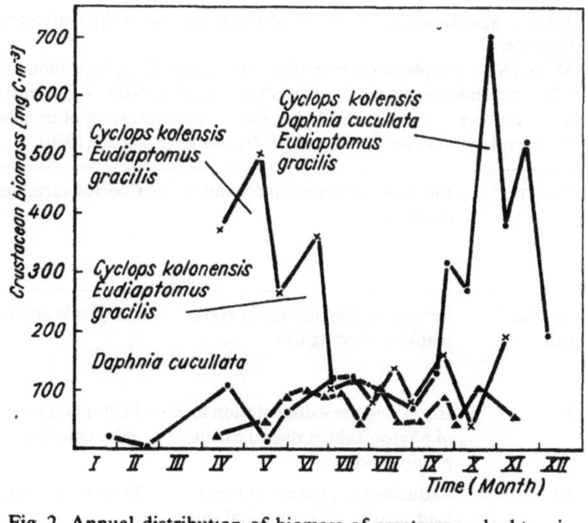

Fig. 2. Annual distribution of biomass of crustacean plankton in Lake Haussee.

1978 (x), 1979 (●) and 1985 (▲) with regard to dominant species. Calculations according to BOTTRELL et al (1976), KASPRZAK (1984), conversion from dry-weight to carbon according to WINBERG et al. (1971) using a conversion factor of 0 5

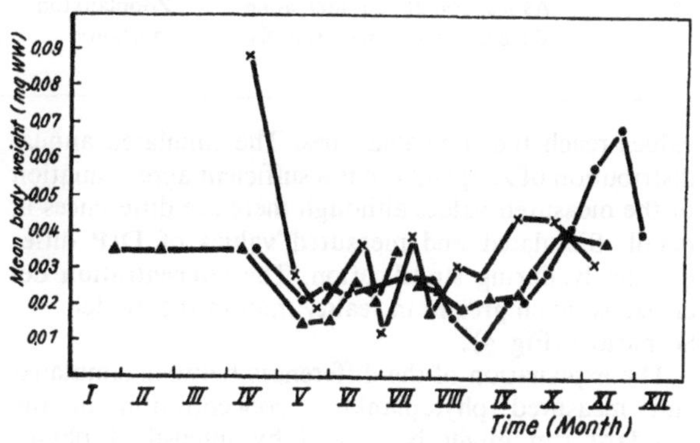

Fig. 3. Annual distribution of mean body weight of crustacean plankton in Lake Haussee.

1978 (x), 1979 (●) and 1985 (▲). Calculations and conversions see Fig. 2.

of the most important part of food-chain with respect to manipulation.

During summer we often found a biomass of about 100 mgC $\cdot$ m^{-3} (Fig. 2). Exceptions are only found in spring and autumn as many planktivorous fish have interrupted their food intake due to low temperatures. Now the crustacean community can realize their production potential as well as a high community filtration rate, although the temperatures do not allow the maximum. Under these conditions "clearwater stages" are also possible in Lake Haussee, as was observed in June 1983. We measured a Secchi depth of 4.5 m and the crustacean biomass was 13 times higher than phytoplankton biomass.

The mean body weight of planktonic crustaceans drops during summer to values of 0.02 mg wet-weight per individual. During the cold season it exceeds 0.03 mg wet-weight per individual clearly (Fig. 3). This can also be the result of heavy fish predation during summer. For that reasons it is advisable to improve the growth conditions of big filter-feeding planktonic crustaceans by means of food-web manipulation to regenerate this part of self-purification potential, thus having a positive effect on water quality.

Results of scenario analyses

The scenario analyses simulate the annual distribution of phytoplankton, zooplankton and DIP assuming a certain set of conditions (Table 2). The results are compared with measured values.

Reference run (Fig. 4):

To begin with we simulated the reference run. It is valid for 1978 at a nutrient load of 1.8 gTP $\cdot$ m^{-2} $\cdot$ a^{-1} and 10 gTN $\cdot$ m^{-2} $\cdot$ a^{-1}. The parameters of zooplankton mortality are rather high (Table 2) and represent a large number of planktivorous fish.

The approximation between simulated and measured values of phytoplankton is good in early spring but noticeable differences become obvious in May and June. In the following period the maxima of the measured

Table 2. Specification and control of the scenarios for the evaluation of the role of various controle mechanisms in the metabolism of Lake Haussee.
MOMIN — zooplankton mortality rate near 0 °C at high biomass density; MOT — constant of temperature dependent mortality rate; VS — net sinking velocity of phytoplankton; PIN, NIN — concentration of dissolved inorganic phosphorus and nitrogen respectively in the water flowing into the mixed layer; P(0) — concentration of dissolved inorganic phosphorus in the water at the beginning of the year. For more explanations see BENNDORF (1979) and RECKNAGEL (1980).

Scenario	Impacts and control mechanism involved	Control variables	Variables or parameters of the model SALMO which are used for the simulation of the scenario	Duration of action of the control mechanism or impact
Standard run	Increasing the biomass of zooplankton-feeding fish	Zooplankton mortality	Parameter MOMIN = $0.03 \cdot d^{-1}$ MOT = $0.012\ °C \cdot d^{-1}$ VS = $0.1\ m \cdot d^{-1}$	Whole year
1	Increasing the sedimentation losses of phytoplankton due to calcite precipitation	Phytoplankton sedimentation	Parameter VS = $0.5\ m \cdot d^{-1}$	1. 5.—20. 6.
2a	Reducing the external nutrient load to 10% of the original value	External nutrient load	Input variable PIN 0.01 NIN 0.01	Whole year
2b	Reducing the internal phosphorus load to 10% of the original value	Phosphorus regeneration from the sediment	Parameter P(0) 0.1	Whole year
3a	As scenario 2a, but increasing the biomass of piscivorous fish	Zooplankton mortality	Parameter MOMIN = $0.0075 \cdot d^{-1}$ MOT = $0.003\ °C \cdot d^{-1}$	Whole year
3b	As scenario 2b, but increasing the biomass of piscivorous fish	Zooplankton mortality	Parameter MOMIN = $0.0075 \cdot d^{-1}$ MOT = $0.003\ °C \cdot d^{-1}$	Whole year

values reach the simulated ones. The simulated annual distribution of zooplankton is a sufficient approximation of the measured values although there are differences in detail. Simulated and measured values of DIP differ drastically during stratification. The concentration decrease is much greater in reality than in the model.

Scenario 1 (Fig. 4):

The explanation of the differences between simulated and measured phytoplankton concentration in the standard run might be caused by intensified phytoplankton sedimentation due to calcite precipitation. However, compared with other surrounding hard-water lakes the calcite precipitation rates in Lake Haussee are smaller (cf. chapter "Discussion"). Nevertheless we increased in scenario 1 the sedimentation rate of phytoplankton for 6 weeks (Table 2) keeping constant all other parameters and variables. Again the spring bloom of phytoplankton is well simulated. Differences are found in April and May but they are much smaller than in the standard run. In the following period the simulated line represents the upper limit of all measured values. In comparison to the standard run there are almost no differences in DIP concentration.

Scenario 2a (Fig. 5):

In 1980 the nutrient loading due to waste-water effluent was interrupted (Table 1). The corresponding decrease of external load of about 90% is reflected in scenario 2a (Table 2). It shows the current trophic state of the lake. The high starting values of nutrients were kept constant as in reality. All other parameters and variables are identical to scenario 1. Compared to that the spring maximum of phytoplankton was lowered to one third and divided into two growth phases. The second peak develops at the beginning of stratification. The differences in summer values are rather small, and also the annual distribution of zooplankton shows almost no deviations. The DIP concentrations decrease a little more as in scenario 1 but remain in the non-limiting range.

Scenario 2b (Fig. 6):

In scenario 2b we have additionally assumed a reduction of internal nutrient load of 90% (Table 2). It reflects the trophic state of Lake Haussee during natural eutrophy as it could have been before the operation of sewage treatment or after complete restoration in future.

Compared with the above cited scenarios the biomass of phytoplankton has decreased clearly. The annual sum amounts to only 50% of that shown in scenario 1. The reason is the simulated, drastically lowered nutrient supply which cannot be observed in reality because of the high internal phosphorus load. Biomass of zooplankton is lower than the measured values especially in spring and autumn.

Scenario 3a (Fig. 5):

Scenario 3a was simulated assuming the same conditions as in scenario 2a excepted the mortality of zooplankton which was decreased to 25% of the standard value (cf. Table 2). It represents the current state of the lake during food-web manipulation. As in scenario 2a the phytoplankton growth remains uneffected in early spring, but at the beginning of stratification remarkable differences become obvious. The above cited second spring maximum (cf. Fig. 5, SC2a) is nearly suppressed and also the concentrations throughout the rest of the year are 50—75% lower. The zooplankton biomass has increased about 100% and the DIP concentration is slightly higher during stratification and autumn.

Scenario 3b (Fig. 6):

The final scenario 3b was simulated under the same conditions as scenario 2b but recognizes a decrease of zooplankton mortality of 75% (Table 2). Consequently its biomass shows a drastic increase. Compared with scenario 2b the annual sum amounts to 350%. As a result the biomass of phytoplankton was lowered to an annual sum of 60%, and the decrease is especially obvious during summer.

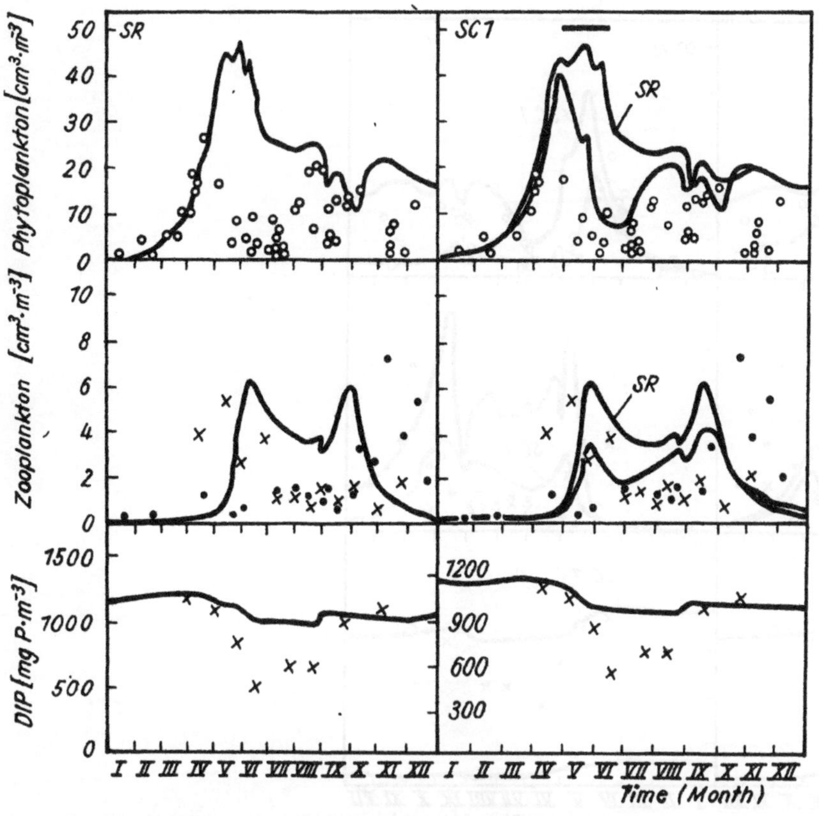

Fig. 4 Standard run (SR) and scenario 1 (SC1), for specification see Table 2.

Horizontal bar — period of increased sedimentation rate due to calcite precipitation from 1 5 —20 6 , o — phytoplankton concentration based on Chl-a measurements from 1978—1985, conversion factor 500, x and ● — measurements from 1978 and 1979 respectively

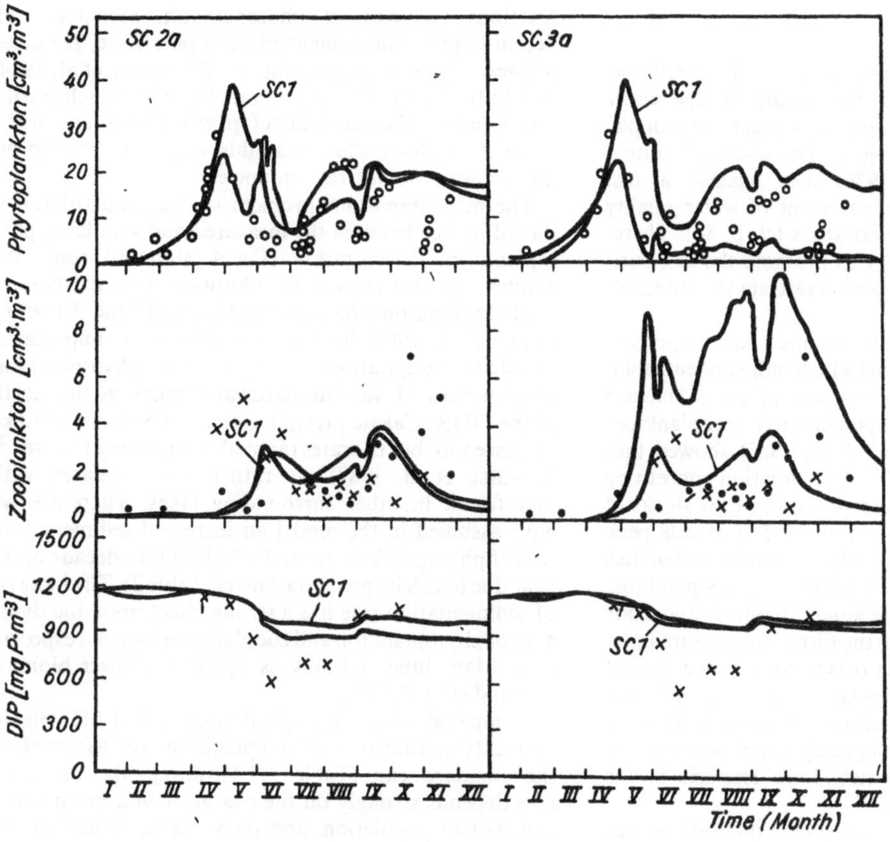

Fig. 5. Scenario 2a (SC2a), scenario 3a (SC3a) and scenario 1 (SC1) as reference

For specification and explanation of symbols see Table 2 and Fig 4 respectively

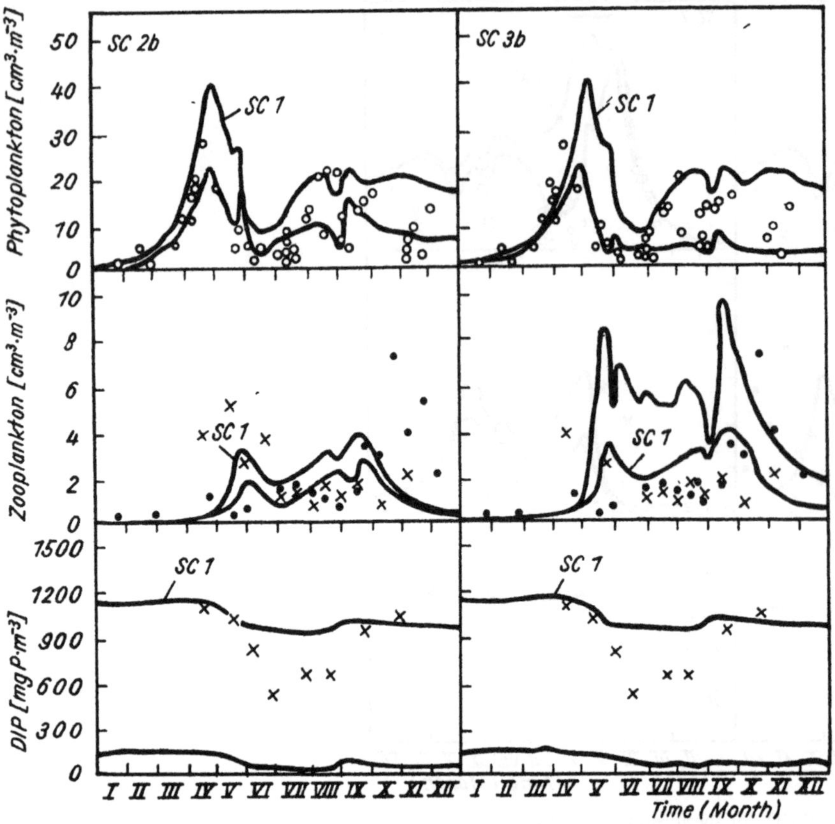

Fig. 6. Scenario 2b (SC 2b), scenario 3b (SC 3b) and scenario 1 (SC 1) as reference.
For specification and explanations of symbols see Table 2 and Fig. 4 respectively

Discussion

Lake Haussee shows still highly eutrophic conditions although the nutrient load due to the sewage water effluent from the municipial treatment plant was stopped five years ago (cf. Table 1, Fig. 1). The so-called "auteutrophication" (GÄCHTER 1976) acts against a fast regeneration. A decisive improvement of water quality cannot be expected in the next years (cf. p. 88). Therefore we looked for a strategy to promote the self-purification potential of the lake and to restrain the auteutrophication.

In the introduction to this volume some important ecological possibilities are cited which influence eutrophication. Analyses of various features of Lake Haussee (hydrography, nutrients, phytoplankton, zooplankton, fish, cf. p. 89, KOSCHEL et al. 1981, 1985) showed that the food-web of the lake is in a bad situation concerning the factors which restrain eutrophication. Low stocks of predacious fish (cf. KOSCHEL et al. 1985, FRANKIE pers. comm.) and the very high stocks of planktivorous fish lead to high mortality of filter-feeding zooplankton. Correspondingly low are the annual biomass and body size (Figs. 2, 3) and therefore the filtration pressure upon phytoplankton too. For this reason we decided for the concept of food-web manipulation (decrease of zooplankton mortality by means of reduction in planktivorous fish). This way is ecologically as well as economically attractive and should turn Lake Haussee into a water of moderate eutrophy.

The aim of the first step is to change the food-web so that zooplankton mortality is lowered to the minimum which can be practically reached. The consequential reactions (decrease of phytoplankton biomass, high Secchi depth, enlargement of eutrophic zone, balanced pH and oxygen concentration, cf. BENNDORF et al. 1984) should be stabilized to regenerate the natural self-purification potential of the lake (cf. p. 89). The results of the scenario analyses show that this way is promising although some details remain unclear.

The most important problem of the standard run is the difference between the measured and simulated phytoplankton concentration (Fig. 4, standard run). We assume for this reason an additional loss mechanism in phytoplankton which acts in May and June. KOSCHEL et al. (1981, 1985) have shown the special importance of calcite precipitation causing losses in phytoplankton communities of various hardwater lakes in the north of the GDR. Calcite precipitation can be shown in Lake Haussee too, but the rates (annual mean 80 mg Ca $\cdot$ m^{-3}, KOSCHEL et al. 1987) are rather low compared with that found in other surrounding lakes. Therefore we have assumed in the model an increased sedimentation rate of phytoplankton from the 13th to 18th decade of the year due to calcite precipitation (cf. Table 2). The increase of sedimentation rate has a strong effect upon the dynamics of phytoplankton and zooplankton biomass especially in May, June and July as significant lower biomass is found (Fig. 4, SC1).

Compared with the original version SALMO higher mortality parameters of zooplankton are assumed in the standard run (Table 2). However, the mortality rate depends strongly on the character of a given water and its fish population, and there are some indications about high losses of zooplankton in Lake Haussee due to fish predation.

Mortality parameters used in the original version of SALMO are based upon investigations of HORN & BENNDORF (1980) carried out in a mesotrophic reservoir but they can also be exceeded in other waters (GLIWICZ et al. 1978, KASPRZAK & RONNEBERGER 1982).

Compared with other less productive lakes the biomass of crustacean plankton is rather low (Fig. 2, KASPRZAK & RONNEBERGER 1982) although a growth limitation due to low food concentration can be excluded (KASPRZAK & RONNEBERGER 1985). Relatively high biomass is found in spring and autumn as many planktivorous fish have interrupted feeding as a consequence of low temperatures. Also the low individual biomass during summer indicates strong fish predation (Fig. 3, BENNDORF et al. 1984). Finally the harvest in small cyprinoid fish has reached a critical value above which typical food-animals disappear because of the "overgrazing effect" (BARTHELMES 1988 and pers. comm.).

There now exist some investigations dealing with nutrition of fish species typical for Middle and North Europe (NYBERG 1979, STENSON 1979, LAMMENS 1984, SCHMIEDS 1984). From these papers it can be concluded that bream (*Abramis brama* L.) and roach (*Rutilus rutilus* L.), which prevail in Lake Haussee, are important plankton predators. In March 1982 we took a number of both species from a commercial catch and investigated the gut content. Up to a length of 24 cm and 17.4 cm respectively we found *Cyclops kolensis* to be the main food organism.

In the standard run as well as in scenario 1 the measured DIP concentration is lower than the simulated one during stratification. Although we have no unambiguous evidence an adsorption of DIP by calcite crystalls at high pH and DIP concentration seems possible.

Scenario 2a (Fig. 5) reflects the conditions after interruption of nutrient loading. Noticeable changes are only found within the phytoplankton whose spring maximum is lowered and the influence of stratification becomes more visible than in the previous scenarios. The DIP concentration remains still high due to internal load.

Under these conditions with respect to the nutrient load the mortality of zooplankton was lowered to 25% in scenario 3a (Table 2) which caused a drastic increase of its biomass (Fig. 5). The spring maximum of phytoplankton remains nearly unaffected because the low temperature prevents development of zooplankton (BOTTRELL et al. 1976). But the second maximum is nearly suppressed and the remaining values are also 50—75% lower compared with those of scenario 2a.

To evaluate this result it is necessary to remember the change in the phytoplankton community of Lake Haussee which moved from small chlorophyceans to filamentous blue-greens. The model recognizes this peculiarity by introducing two functional groups of algae. One represents small and fast growing phytoplankters from the type of "*Stephanodiscus*" and "*Scenedesmus*" and the other one filamentous blue-greens of the type of "*Oscillatoria*". According to DAWIDOWICZ et al. (1988) the maximum filtration rate of *Daphnia magna* is lowered to 50% when the concentration of those blue-greens is only 18 $\mu g \cdot l^{-1}$. The limit is reached in Lake Haussee (cf. Fig. 1) but the model does not reflect the interaction in this manner. Therefore the simulated biomass of phytoplankton in scenario 3a can be slightly too low.

Scenario (Fig. 6, SC 2b) was simulated assuming a 90% reduction of internal and external nutrient load. The phytoplankton biomass decreases in an order of magnitude which is similar to that after food-web manipulation (cf. Fig. 5, SC 3a).

Although according to ANWAND (1973) Lake Haussee must be classified as pike-perch lake with natural recriutment of the species (FRANKIE, pers. comm.) notwithstanding the bad water quality, stocking of one-summer-old fishes is problematic. According to LADIGES & VOIGT (1979) and BARTHELMES (1981) pike-perch prefers smelt (*Osmerus eperlanus* L.), bleak (*Alburnus alburnus* L.), and white bream (*Blicca bjoerkna* L.) and it may be that this feeding habit promote the dominating bream stocks of Lake Haussee. On the other hand NIKOLSKI (1957) refers to an adaptable prey selection of pike-perch which is clearly influenced by the structure of small-fish population. However, pike-perch introduction was most successful in small eutrophic lakes (LEHTONEN 1984).

It is well known that eutrophication is no simple interaction between nutrients and algae but leads to a complex change of the whole ecosystem and in many cases the results are not completely understood. Concerning food-web manipulation LAMPERT (1983) speaks therefore about the close relationship of success and disappointment bacause of the permanent tendency of phytoplankton to overcome the artifical production limitation (cf. *p.* 87). However, due to drastic reduction of external nutrient load an prerequisite is provided to diminish this trend. Moreover some conditions are established for the functioning of important ecological side-effects of food-web manipulation. KOSCHEL et al. (1983, 1987) have shown that pelagic calcite precipitation and the connected decrease of primary production are especially efficient in mesotrophic lakes. Because food-web manipulation leads also to limitation of phytoplankton growth a stimulation of calcite precipitation can be assumed which influences the metabolism of the lake and acts additionally against eutrophication. Finally an intensified development of phyto- and zoobenthos can happen leading to lower turnover rates of nutrients (LEAH et al. 1980), and it can be hoped that after some years of food-web manipulation Lake Haussee will maintain a lower level of eutrophication without artificial stocking of pikeperch.

Zusammenfassung

Der Haussee (Feldberg, DDR) wurde durch die Einleitung der Abwässer einer kommunalen Kläranlage stark eutrophiert. Seit 1980 ist die Einleitung unterbrochen, aber der trophische Zustand des Gewässers änderte sich kaum. Ursache dafür sind die große Menge gespeicherter Nährstoffe und die hohe interne Belastung. Untersuchungen zur Hydrologie und Hydrochemie sowie über die Nahrungskette zeigten, daß das Selbstreinigungspotential des Sees nicht mehr voll entwickelt ist. Wir haben deshalb den Effekt der Nahrungskettenmanipulation auf die Wasserqualität des Haussees theoretisch unter Verwendung des ökologischen Seenmodells SALMO untersucht.

Die Reduktion der Zooplanktonmortalität, so hat sich gezeigt, führt zu einem drastischen Rückgang der Phytoplanktonbiomasse. Der zweite Teil des Frühjahrsmaximums wird nahezu vollständig unterdrückt, und im weiteren Jahresverlauf liegen die Werte um 50—75% niedriger.

Wir vermuten, daß zwischen der Nahrungskettenmanipulation und der Kalzitfällung, einem anderen wichtigen Selbstreinigungsmechanismus von Hartwasserseen, eine Verbindung besteht. Die pelagische Kalzitfällung und die damit verbundene Einschränkung der Primärproduktion wirken am effektivsten in mesotrophen Seen.

Da die Nahrungskettenmanipulation ebenfalls zu einer Limitation des Phytoplanktonwachstums führt, scheint die Anregung der Kalzitfällung möglich zu sein.

References

ANDERSSON, G., BERGGREN, H., CRONBERG, G., & GELIN, C (1978): Effects of planktivorous and benthivorous fish on organısms and water chemistry in eutrophic lakes. Hydrobiologia **59**, 9—15.

ANWAND, K. (1973): Gewässerverzeichnis der Seen- und Flußfischerei der DDR. Inst. Binnenfisch. Berlin-Friedrichshagen, pp. 1—100.

BARTHELMES, D. (1978): Hypothese über einen Zusammenhang Raubfischwirtschaft — Fischertrag — Wasserqualitat beı hoher Trophie. Z. Binnenfisch DDR **25**, 274—277.

— (1981): Hydrobiologische Grundlagen der Binnenfischereı. Jena, pp. 1—250.

BENNDORF, J. (1979)· Kausalanalyse, theoretische Synthese und Simulation des Eutrophierungsprozesses in stehenden und gestauten Gewässern Diss B, Techn. Univ. Dresden, pp. 1—165.

— (1985): Dynamic ecologıcal models as a decisionmakıng aid in eutrophicatıon control. Wıad. Ekol. **31**, 335—349 (in Polish, with English summary and figures).

— & RECKNAGEL, F. (1978): Entwicklungsstand von Modellen zur Vorhersage des Eutrophierungsprozesses in stehenden und gestauten Gewässern. Mitt. Chem. Ges. DDR, Beih. **20**, 50—51.

— — (1982): Problems of application of the ecological model SALMO to lakes and reservoirs. Ecol. Model. **17**, 129—145.

— KNESCHKE, H., KOSSATZ, K., & PENZ, E. (1984): Manipulation of the pelagic food web by stocking with predacious fishes. Int. Rev. ges. Hydrobiol. **69**, 407—428.

— KOSCHEL, R., & RECKNAGEL, F. (1985): The pelagic zone of Lake Stechlin. An approach to a theoretical model. In: S. J. CASPER (ed.), Lake Stechlin. A temperate oligotrophic lake. Dordrecht-Boston-Lancaster, pp. 433—453.

— SCHULZ, H., BENNDORF, A., UNGER, R., PENZ, E., KNESCHKE, H., KOSSATZ, K., DUMKE, R., HORNIG, U., KRUSPE, R., & REICHEL, S. (1988): Food-web manipulation by enhancement of piscivorous fish stocks: Long-term effects in the hypertrophic Bautzen reservoir. *Limnologıca* (Berlin) **19** (1), 97—110

BJÖRK, S., FLEISCHER, S., GRANELI, W., BENGTSSON, L., LEONARDSON, L., ANDERSSON, G., & RIPL, W. (1979): Lake management, studies and results at the Institute of Limnology in Lund. Arch. Hydrobiol., Beih. **13**, 31—55.

BOTTRELL, H. H., DUNCAN, A., GLIWICZ, Z. M., GYGIEREK, E., HERZIG, A., HILLBRICHT-ILKOWSKA, A., KURASAWA, A., LARSSON, P., & WEGLENSKA, T. (1976): A review of some problems in zooplankton production studies. Norw. J. Zool. **24**, 419—456.

BRANDL, Z., & FERNANDO, C. H. (1975): Food consumption and utilization in two freshwater cyclopoid copepods (*Mesocyclops edax* and *Cyclops vicinus*). Int. Rev. ges. Hydrobiol. **60**, 471—494.

BRIAND, F., & MCCAULEY, E. (1978): Cybernetic mechanisms in lake plankton systems, how to controle undesırable algae. Nature **273**, 227—230.

BROOKS, J. L., & DODSON, S. I. (1965): Predation body size and composition of plankton. Science **150**, 28—35.

DAWIDOWICZ, P., GLIWICZ, Z. M., & GULATI, R. D. (1988): Can *Daphnia* prevent a blue-green algal bloom in hypertrophıc Lakes? A laboratory test. *Limnologıca* (Berlın) **19** (1), 21—26.

DUNCAN, A. (1975): The importance of zooplankton ın the ecology of reservoirs. Proc. Symp. "The effects of storage on water quality", Reading Univ., Water Res. Cent., Medanmanham Lab., England, pp. 247—272.

EDMONDSON, W. T., & LITT, A. H. (1982): *Daphnia* ın Lake Washington. Limnol. Oceanogr. **27**, 272—293.

— & ABELLA, S E B (1987): Unplanned biomanipulation in Lake Washington. *Lımnologıca* (Berlin) **19** (1), 73—79.

FEDORENKO, A. Y. (1975): Instar and specıes specıfic diet ın two species of *Chaoborus*. Limnol. Oceanogr **20**, 238—249.

FOTT, J. (1975): Seasonal succession of phytoplankton ın the fishpond Smyslov near Blatna, Czechoslowakıa. Arch Hydrobıol Suppl. **46**, 259—279.

— PECHAR, L., & PRAZAKOVA, M. (1980): Fish as a factor controlling water qualıty in ponds. Developm. in Hydrobiol. **2**, 255—261.

GÄCHTER, R. (1976): Die Tiefenwasserableitung, ein Weg zur Seensanierung. Schweiz. Z Hydrol. **38**, 1—28

GLIWICZ, Z. M., HILLBRICHT-ILKOWSKA, A., & WEGLENSKA, T. (1978): Contribution of fish and invertebrate predation to the elımination of zooplankton biomass in two Polish lakes. Verh. Int. Verein. Limnol. **20**, 1007—1011.

GOAD, J. A. (1984): A biomanipulation experiment in Green Lake, Seattle, Washington. Arch. Hydrobıol. **102**, 137—153.

HENRIKSON, L., NYMANN, H. G., OSCARSON, H. G., & STENSON, J. A. E. (1980): Trophic changes without changes in external nutrıent load Hydrobiologia **68**, 257—263.

HESSEN, D. O (1983): Introduction of pike *Esox lucius* to a small pond, effects on planktivorous fish, zooplankton and algae Fauna **36**, 119—124

HORN, W. (1981): Phytoplankton losses due to zooplankton grazing ın a drıngkıng water reservoir. Int. Rev ges. Hydrobıol. **22**, 787—801.

— & BENNDORF, J. (1980) Field investıgatıon and model simulation of the dynamics of zooplankton populations in fresh waters. Int. Rev. ges Hydrobiol. **65**, 209—222.

HRBÁČEK, J. (1958): Typologie und Produktivität der teichartıgen Gewässer. Verh. Int. Verein. Limnol. **13**, 394—399.

— DVOŘAKOVA, M., KOŘINEK, V., & PROCHÁZKOVÁ, L. (1961): Demonstration of the effect of fish stock on the species composition of zooplankton and the ıntensity of metabolism of the whole plankton associatıon. Verh. Int. Verein. Limnol. **14**, 192—195.

— DESORTOVA, B., & POPOVSKY, J. (1977): The influence of the fish stock on the phosphorus chlorophyll ratio. Verh. Int. Verein. Limnol. **20**, 1624—1628.

KAJAK, Z., & RYBAK, J. (1979): The feeding of *Chaoborus flavicans* MEIGEN (Diptera, Chaoboridae) and its predation on lake zooplankton. Int. Rev. ges. Hydrobiol. **64**, 361—378.

KARABIN, A. (1974): Studies on the predatory role of the cladoceran *Leptodora kindtii* (FOCKE) in secondary production ın two lakes with different trophy. Ecol. Pol. **22**, 295—310.

KASPRZAK, P. (1984): Bestimmung des Körperkohlenstoffs von Planktoncrustaceen. *Limnologıca* (Berlin) **15** (1), 191—194.

— & RONNEBERGER, D. (1982): Vergleichende Untersuchungen zur Struktur und Dynamik des Zooplanktons im Stechlinsee, Nehmıtzsee und Haussee 1978/79. *Limnologica* (Berlin) **14** (2), 263—295.

— — (1985). The secondary production. In: S. J. CASPER (ed.), Lake Stechlin, a temperate oligotrophic lake. Dordrecht-Boston-Lancaster, pp. 323—345.

KERFOOT, W. C. (1977): Implication of copepod predation. Limnol. Oceanogr. **22**, 316—325.

KOSCHEL, R., HAUBOLD, G., KASPRZAK, P., KÜCHLER, L., PROFT, G., & RONNEBERGER, D. (1981): Eine lımnologische Zustandsanalyse des Feldberger Haussees. Acta hydrochim. et hydrobiol. **9**, 255—279.

— BENNDORF, R., PROFT, G., & RECKNAGEL, R. (1983): Calcite precipitation as a natural control mechanism of eutrophication. Arch. Hydrobiol. **98**, 390—408.

— CASPER, S. J., DOLGENER, W., FESTERSEN, O., KASPRZAK, P., KRAUSCH, H.-D., PÄCHNARTZ, M., PENZ, E., PROFT, G., & RONNEBERGER, D. (1985): Das Feldberger Seengebiet. Ein Beitrag zum Schutz und zur optimalen Nutzung der Feldberger Seen. Natur Umwelt Bez Neubrandenburg **3**, 1—96.

— PROFT, G., & RAIDT, H. (1987): Autochthone Kalkfällung in Hartwasserseen der Mecklenburger Seenplatte. *Limnologica* (Berlin) **18** (2), 317—338.

LADIGES, W., & VOIGT, D. (1979): Die Süßwasserfische Europas. Hamburg, Berlin, pp. 1—3 α0.

LAMMENS, H. R. R. (1984): A comparison between the feedıng of white bream (*Blicca bjòrkna*) and bream (*Abramıs brama*). Verh. Int. Verein Limnol. **22**, 886—890.

— (1987): Trophıc interactıons in the hypertrophic lake Tjeukemeer: Top-down and bottom-up effects in relation to hydrology, predatıon an bıoturbatıon durıng the perıod 1974—1985. *Limnologica* (Berlin) **19** (1), 81—85.

LAMPERT, W (1983) Bıomanıpulatıon — eine neue Chance zur Seensanıerung? Bıologıe ın unserer Zeit **13**, 86—97

— & SCHOBER, U. (1978): Das regelmäßige Auftreten von Frühjahrsalgenmaxima und „Klarwasserstadium" im Bodensee als Folge klimatischer Bedingungen und Wechselwirkungen zwischen Zoo- und Phytoplankton. Arch. Hydrobiol. **82**, 364—386.

LEAH, R. T., MOSS, B., & FORREST, D. E. (1980): The role of predation in causing major changes in the limnology of a hyper-eutrophic lake. Int. Rev. ges. Hydrobiol. **65**, 223—247.

LEHTONEN, H., MIIN, A., & FRISK, T (1984): Natural occurrence of pike-perch (*Stizostedion lucioperca* L.) and success of introductions in relation to water quality and lake area in Finland. Aqua Fennica **14**, 189—196.

MCCAULEY, E., & BRIAND, F. (1979): Zooplankton grazing and phytoplankton species richness, field test of the predation hypothesis. Limnol. Oceanogr. **24**, 243—253.

MURTAUGH, P A. (1981): Selective predation by *Neomysis mercedis*. Limnol. Oceanogr. **26**, 445—453.

NIKOLSKI, G. W. (1957): Spezielle Fischkunde. Berlin, pp. 1—632.

NYBERG, P. (1979): Production and food consumption of perch, *Perca fluviatilis* L., in two Swedish forst lakes Inst. Freshwat. Res. Drottningholm, Rep. No **58**, 140—157.

O'BRION, W. J. (1979): The predator-prey interaction of planktivorous fish and zooplankton. Amer. Sci **67**, 572—581.

OHLE, W (1934): Chemische und physikalische Untersuchungen Norddeutscher Seen. Arch. Hydrobiol. **26**, 386—464 and 583—658.

RECKNAGEL, F. (1980): Systemtechnische Prozedur zur Modellierung und Simulation von Eutrophierungsprozessen in stehenden und gestauten Gewässern. Diss. A, Techn. Univ. Dresden, pp. 1—155.

— (1985): Analysis of structural stability of aquatic ecosystems as an aid for ecosystem control. Ecol. Model. **27**, 221—234.

REINERTSEN, H., & OLSEN, Y. (1984): Effects of fish elimination on the phytoplankton community of a eutrophic lake. Verh. Int. Verein. Limnol **22**, 649—657.

SCHMIEDS, U. J. (1984): Ökologische Studien an Fischen im Schierenbach, einem norddeutschen Seenausfluß. II. Die Nahrung der häufigsten Cypriniden. Faun.-ökol. Mitt. **5**, 217—226.

SHAPIRO, J., & WRIGHT, D. (1984): Lake restoration by biomanipulation, Round Lake, Minnesota, the first two years. Freshwat. Biol. **14**, 371—383.

— LAMARRA, V., & LYNCH, M. (1975): Biomanipulation, an ecosystem approach to lake restoration. Proc. Symp. "Water quality management through biological control", Univ. Florida Gainesville, P. L. BREZONIK & J. L. FOX (eds.), pp. 85—96.

STEEL, J. A. (1975): The management of Thames valley reservoirs. Proc. Symp. "The effects of storage on water quality", Reading Univ., Water Res. Cent., Medmenham Lab., England, pp. 371 to 419.

STENSON J. A. E. (1978): Relation between vertebrate and invertebrate zooplankton predators in some arctic lakes. Astarte **11**, 21—26.

— (1979): Predator — prey relations between fish and invertebrate prey in some forest lakes. Inst. Freshwat. Res. Drottningholm Rep. No. **58**, 166—183

THIENEMANN, A. (1925): *Mysis relicta*. Z. Morphol. Ökol. Tiere **3**, 389—440.

UHLMANN, D. (1954): Zur Kenntnis der natürlichen Nahrung von *Daphnia magna* und *Daphnia pulex* Z. Fischerei Hilfswiss. **3**, 449—478.

WINBERG, G. G., MANN, K. H., TALLING, J. F., GOLTERMAN, H. L., & BLAZKA, P (1971): Symbols, units and conversion factors in studies of freshwater productivity. IBP Control Office, London, pp. 1—23.

ZARET, T. M. (1980): Predation and freshwater communities. New Haven-London, pp. 1—179.

Received January 20, 1987

Address of the authors:

Dr. PETER KASPRZAK
Dr. RAINER KOSCHEL
Akademie der Wissenschaften der DDR
Zentralinstitut für Mikrobiologie und
Experimentelle Therapie Jena
Abteilung Limnologie
Beutenbergstr. 11
Jena
DDR-6900

Doz. Dr. JÜRGEN BENNDORF
Doz. Dr. FRIEDER RECKNAGEL
Technische Universität Dresden
Sektion Wasserwesen
Bereich Hydrobiologie
Mommsenstr. 13
Dresden
DDR-8027

Limnologica (Berlin) 19 (1), 97—110, Juli 1988

Technische Universitat Dresden, Sektion Wasserwesen, Bereich Hydrobiologie, Dresden, DDR

Food-web Manipulation by Enhancement of Piscivorous Fish Stocks: Long-term Effects in the Hypertrophic Bautzen Reservoir

By JÜRGEN BENNDORF, HEINZ SCHULTZ, ANGELA BENNDORF, RAINER UNGER, ECKHARD PENZ, HELMUT KNESCHKE, KATHRIN KOSSATZ, ROGER DUMKE, UWE HORNIG, RAINER KRUSPE & SUSANNE REICHEL

With 15 Figures and 4 Tables

Key words: Eutrophication control; Biomanipulation, Long-term stability; Fish; Zooplankton; Phytoplankton: Bacteria; Phosphorus

Abstract

The enhancement of piscivorous fish by stocking with pond-raised O+ pikeperch (*Stizostedion lucioperca*) and catch restrictions for pikeperch and pike (*Esox lucius*) were found to be an appropriate method of controlling planktivorous fish to a moderate density in the hypertrophic Bautzen reservoir (A = 533 ha, $\bar{z}$ = 7 4 m). The response of the whole ecosystem during the first five years of the biomanipulation period (1981—1985) is compared with the pre-biomanipulation period (1977—1980). This comparison shows that the principial mechanisms of biomanipulation in a hypertrophic water heavily loaded external phosphorus sources are almost the same as in waters having lower phosphorus loads. But the probability of an improvement of the water quality by biomanipulation decreases with increasing P-load due to the high number of "degrees of freedom" of a nutrient-saturated ecosystem. Some responses of the ecosystem to the food-web manipulation show remarkable instability during five years (e g. steady decrease of the mean individual body size of the herbivorous crustaceans, strong up- and downward trends in the abundance of *Chaoborus* and strong fluctuations in the abundance of *Leptodora*, bistability of the structure of the summer phytoplankton). These instabilities are regarded to be the consequence of the stochastic character of the meteorological conditions as well as the result of the time lag of the feedback mechanisms within the ecosystem. The conclusion is drawn that (1) maximum possible long-term stability of food-web manipulation can be achieved by controlling the fish community in such a way that a stable moderate stock of planktivorous fish is guaranteed rather than by exterminating these fish (e.g. by poisoning) and (2) effective eutrophication control of water bodies having a high external phosphorus load seems to be possible only by combining food-web manipulation with other control measures (e.g. reduction of the external P-load, destratification).

Introduction

Great success has been achieved in research on structuring pelagic food-webs since the stimulating work of HRBÁČEK et al. (1961) revealed the enormous influence of fish on the other elements of aquatic ecosystems. A great part of this knowledge has been gathered by various experiments in "biomanipulation" (SHAPIRO et al. 1975) which have been carried out during the last decade [overview see BENNDORF (1988): Table 2]. But if biomanipulation should be used as a stable measure of eutrophication control, numerous problems remain unsolved, for instance:

(1) Development of inedible algae,
(2) Long-term balance of predation and competition in all communities,
(3) Role of spatial inhomogeneity in the physical and chemical conditions,
(4) Role of various nutrient loads,
(5) Appropriate methods of controlling the food-web [see BENNDORF (1987) for a more detailed explanation of these five problems].

The results of the biomanipulation experiments hitherto published do not offer sufficient answers to many of these questions because (a) almost all of the experimental waters are rather small, (b) only in exceptional cases were the experiments extended to more than two or three years, (c) only a few comparative investigations are available of water bodies having extremly different phosphorus loads, and (d) the majority of the biomanipulation experiments started with the drastic measure of poisoning all fish.

The aim of this study is to contribute to the solution of the open problems by a long-term investigation of a relatively large water body, which is subjected to an extremly high phosphorus load. Furthermore, enhancement of predacious fish by different methods is the only measure for controlling the food-web.

Site description

Bautzen reservoir (Fig. 1) is a man-made lake about 70 km north-east of Dresden (GDR). Some important characteristics of this water body are given in Table 1. Due to the lack of horizontal protection and the circle-shaped surface the water usually exhibits a thermal stratification for only a few weeks during summer. The main water quality problem consists in excessive eutrophication (low Secchi depth, algal blooms, extreme variations in the oxygen concentration and pH). Bautzen reservoir is a typical multi-purpose reservoir (cooling water supply for a power plant, water supply for fish ponds situated downstream, flood control, recreation). The intensive recreational use made of the reservoir (e.g. camping, boating, surfing, angling, swimming) makes the highest demand on the water quality.

Using the scenario technique and a dynamic ecological model (SALMO, cf. BENNDORF 1979, RECKNAGEL 1984) biomanipulation was found to be the most promising

Table 1. Basic characteristics of the Bautzen reservoir

Volume	39 2 · 10^6 m^3
Surface	533 ha
Mean depth	7.4 m
Mean theoretical retention time of the water	193 d
Exposition to wind	Extremely exposed
P-loading	4.1 to 15.5 gP · m^{-2} · $year^{-1}$ (total phosphorus; see Fig. 14)
Eutrophication degree	Hypertrophic

Fig. 1. View from the inlet section to the central part of the Bautzen reservoir.

strategy for eutrophication control in the Bautzen reservoir, and the implementation of this strategy began in 1977 with the Dresden University of Technology, the state authorities and the sporting fishermen's union of the GDR working in cooperation. This cooperation is a decisive organizational prerequisite in biomanipulation.

Methods

Samples for analyzing the physical and chemical criteria and the plankton community were taken in three-weekly intervals as vertical series (2.5 m spacing, 2 l-Ruttner sampler). In some years sampling took place at three stations (cf. HORNIG & BENNDORF 1985), but for the long-term comparisons discussed in this paper, values of station A (in the deepest part of the water body) are used exclusively.

Bacteria and detritus were counted in subsamples of the vertical series. These subsamples were preserved immediately in 4% formalin. For filtration (Sartorius membrane filters 0.2 µm) 1 ml of these subsamples was diluted with 9 ml bacteria-free water in order to achieve an even distribution on the filter. Counting was carried out, after staining with carbolerythrosine, using an phase-contrast/interference microscope (Zeiß, Jena). The same instrumentation was employed to measure the cell size, so that the calculation of the bacterial biovolume was possible. Compared with the epifluorescence technique, better results (higher numbers, higher precision) were achieved with the phase-contrast/interference method. Detritus particles were counted in small chambers (volumes 4 mm³ and 100 mm³, respectively) depending on size and number of the particles. The calculation of the detritus volume was done by assigning all particles to seven size classes (3 to 5 µm, 5 to 10 µm, ..., 30 to 35 µm). The estimate of the volume V of the detritus aggregates is achieved by equ. (1).

$$V = \frac{\pi}{18} \cdot d^3 \qquad (1)$$

with d — median value of the size class concerned.

Phytoplankton samples were examined live about 2 h after collection and, after preservation with Lugols solution, counted by the UTERMÖHL (1958) sedimentation technique. The values were converted into biovolumes by comparison with simple geometrical bodies. The biovolume of *Microcystis* colonies was calculated by an analogous procedure as described for detritus.

Zooplankton sampling was carried out in different ways. Ciliates were counted in the phytoplankton samples. Rotifers as well as small and medium-sized crustaceans were collected by a 2 l-Ruttner sampler and concentrated using a net, which had a mesh size (diagonal) of 50 µm. Large zooplankters (*Leptodora*, *Chaoborus*, adult *Daphnia*, *Cyclops* and *Eudiaptomus*) were caught by vertical hauls. Two different nets were employed (diagonal mesh size 500 µm and 900 µm, respectively, opening diameter 30 cm). The samples were immediately preserved in a mixture of formol and saccharose. Counts were performed on 5 to 10 subsamples, depending on the abundance of the animals, using a KOLKWITZ chamber. When abundances were low, total counts were performed after the whole sample had been sedimented in an UTERMÖHL chamber. A microscope with a TV screen attachment permitted the rapid assignment of all individuals to size classes by means of a simple ruler. The class intervals (body length) were usually 50 µm. Empirical relationships between body length and body volume (BOTTRELL et al. 1976, HORN 1978) were used to calculate the biovolumes, the median value of the size class concerned being used for this purpose.

In addition to the vertical hauls, *Chaoborus* larvae were also collected from the sediment by means of an Ekman-sampler. Usually five samples were taken around station A at each sampling date.

A combination of gill net catching and hydroacoustic counting was employed to estimate the density and composition of the fish community (SCHULTZ 1983, 1988).

Control and development of the fish community

In the two years after the initial impoundment of the Bautzen reservoir (1974 and 1975), a numerous pike population (*Esox lucius*) developed in a natural way. Fastgrowing pike occurred in large quantities in 1976. At this time (May 1976) sport fishing started at the reservoir which led to a steady decline in the pike population. Rough estimates showed that in 1976 to 1978 about

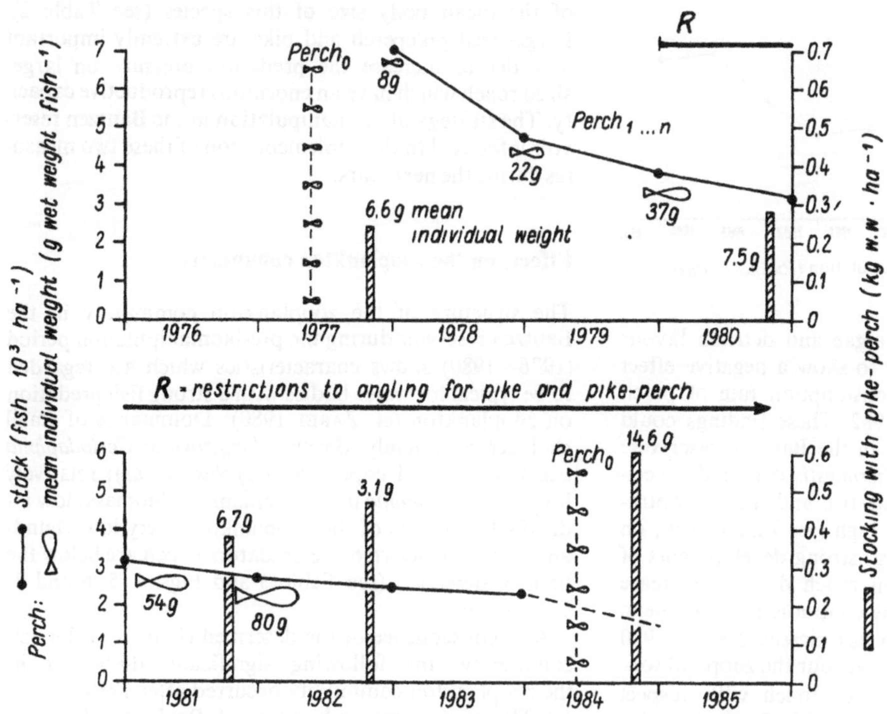

Fig. 2. Measures to increase the abundance of piscivorous fish pikeperch (*Stizostedion luciperca*), pike (*Esox lucius*) and large perch (*Perca fluviatilis*). Development of the perch population. (Perch smaller than about 50 g is by preference zooplanktivorous)

2,000 to 3,000 pike were caught per year. The consequence of this overfishing was the maintenance of an abundant stock of small zooplankton-eating perch (*Perca fluviatilis*) for several years. Most of these perch belong to the year-class 1977 (SCHULTZ 1983).

With the beginning of biomanipulation in 1977, each year (with the exception of 1978, 1979 and 1983) 20,000 to 80,000 pond-raised O+ pikeperch were introduced in autumn, and restrictions regarding the catch of predatory fish were imposed on sporting fishermen after 1979 (Fig. 2). These restrictions involve the following regulations: Only one specimen of predatory fish is allowed to take per day and the minimum size must be 60 cm for pike and pikeperch; the use of fish as baits is prohibited, exclusively artificial lures are allowed. One of the aims of our biomanipulation experiment is to use mainly biological control by optimizing the stocks of predacious fish. As a result of these measures the remaining pike population was stabilized and an increasing population of pikeperch (*Stizostedion lucioperca*) was established (Table 2). It should be stressed that this was only possible by combining both the control possibilities (stocking **and** catch restrictions).

It is most likely that these measures have reduced the abundance of the perch population in two different ways: (1) Mortality of the year-class 1977 due to pike predation, and (2) low recruitment of the perch population due to predation by pikeperch. As Fig. 2 shows, only moderately large numbers of perch existed in 1981, i.e. when biomanipulation had become effective. At the same time the average body weight of perch increased (Fig. 2). Due to the decreasing number of perch and their low consumption rates (see KÖPKE et al. 1988) the predation pressure on zooplankton was essentially reduced in 1981 and 1982. In the following years consumption of zooplankton by perch was further diminshed by a stronger shift from predation on zooplankton to predation on fish. Thus, a well-developed stock of perch guarantees the fast reduction of the enormously numerous young-of-the-year of all fish species (including its own species). This latter mechanism seems to be a key factor in stabilising low fish predation on zooplankton and it has functioned in the Bautzen reservoir for a few years. In 1984 a considerable proportion of the perch population died in a natural way after spawning due to a disease and/or other reasons. As a consequence of this decrease in the perch population, large numbers of fish fry were observed in 1984 (perch) and 1985 (roach). These young fish were reduced to a low abundance by pikeperch during 1984 and 1985; but this reduction becomes effective later in the year than perch predation on fry and, thus, cannot minimize the strong fish fry predation on zooplankton during summer.

Table 2 Development of the population of pikeperch (*Stizostedion lucioperca*) Time refers to the end of the previous and the beginning of the following year. Numbers of fish per hectare and individual body weight are given without the O + group. See Fig. 2 for information about stocking with O+pikeperch

Period	Time	Fish per hectare	Individual body weight (g)	
			Mean value	Maximum
Prebiomanipulation	77/78	5	188	200
	78/79	11	254	500
	79/80	9	597	1000
	80/81	10	675	1400
Biomanipulation	81/82	19	625	2200
	82/83	26	666	2800
	83/84	50	508	3500
	84/85	41	789	5100
	85/86	51	691	4800

During the first years of biomanipulation the roach population (*Rutilus rutilus*) scarcely increased in number (Fig. 3). But in the eigthies roach became more and more abundant. This succession from perch to roach dominance is well documented for eutrophic waters (PERSSON 1983 a). PERSSON (1983 b) and LESSMARK (1983) have found that

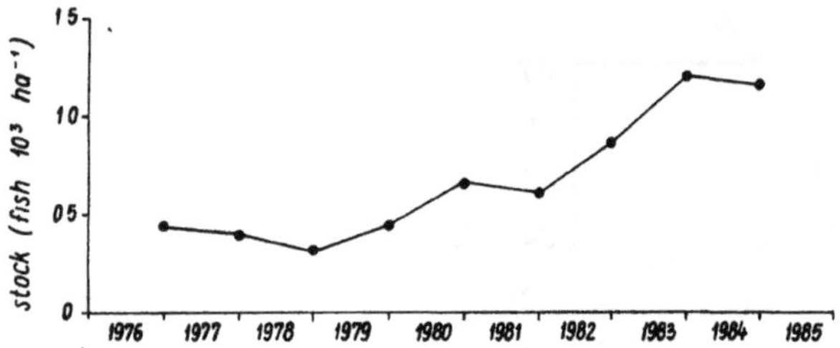

Fig. 3. Development of the roach population (*Rutilus rutilus*)

high amounts of blue-green algae and detritus favour roach. PENIG (1983) was able to show a negative effect of blue-green algae on the consumption rate of perch in the Bautzen reservoir in 1982. These findings could explain the increase of roach in the Bautzen reservoir: Extremly high densities of *Microcystis* occurred especially in 1982 and 1983 (see Fig. 10), and the concentration of detritus is also rather high (see Fig. 8). But, on the other hand, there were also strong developments of blue-greens before 1980 when roach did not increase in number. In addition to this contradiction, the mean body size of the zooplankton has increased since 1980 (see Figs. 5 and 6) which should favour the zooplankton-eating size-classes of perch over roach with respect to competition for food (LESSMARK 1983). But it must also be taken into consideration that more food (zooplankton) was left for roach as a consequence of the decreasing feeding pressure of perch on zooplankton. Thus, at this stage of the experiment there is no sufficient explanation of the observed development of the perch-roach-relationship.

The only promising way to stop the increase of the roach population seems to be (1) to control blue-greens by artificial destratification (SHAPIRO 1984), and (2) the further increase of the pikeperch stock as well as of the mean body size of this species (see Table 2). Large-sized pikeperch and pike are extremly important in order to increase the predation pressure on large-sized roach which have an enormous reproductive capacity. The strategy of biomanipulation in the Bautzen reservoir is focused to the implementation of these two measures during the next years.

Effects on the zooplankton community

The structure of the zooplankton community in the Bautzen reservoir during the pre-biomanipulation period (1976—1980) shows characteristics which are regarded to be typical for water bodies with a strong fish predation on zooplankton (cf. ZARET 1980): Dominance of small cladocerans (mainly *Bosmina longirostris*; *Ceriodaphnia quadrangula*) and copepodes (*Cyclus vicinus*); relatively low share of *Daphnia* in the zooplankton biomass; low individual body size of the zooplankton; very low abundances of large invertebrate predators which are below the limit of detection (see Table 3 and Figs. 4, 5, 6 and 7: 1976 to 1980).

As a consequence of the described changes in the fish community, the following significant alterations in the zooplankton community occurred after 1980:

(1) The mean summer biomass of *Daphnia galeata* increased from 1.966 mg wet weight · 1^{-1} (1976 to 1980) to 4.748 mg w.w. · 1^{-1} (1981 to 1985), thus having a relative share in the total zooplankton biomass of 62% during biomanipulation (Table 3, Fig. 4). It must be emphasized that the total biomass of the herbivorous zooplankton did not increase during the whole investigation period (summer averages: 7.516 mg w.w · 1^{-1} from 1976 to 1980; 7.609 mg w.w · 1^{-1} from 1981 to 1985; Table 3, Fig. 4).

Table 3. Characteristics of the zooplankton community in the Bautzen reservoir [values from May to October; *n* (per season) = 9 ... 12]

Criteria	Pre-biomanipulation period					Biomanipulation period				
	1976	1977	1978	1979	1980	1981	1982	1983	1984	1985
Mean biomass of herbivores (mg w.w. · 1^{-1})	11.5	3.8	7.7	10.0	4.6	6.6	7.4	8.2	7.0	8.8
Maximum biomass of herbivores (mg w.w. · 1^{-1})	58.5	11.9	16.6	17.9	16.0	13.9	13.4	22.2	18.3	19.1
Mean share of *Daphnia* in the biomass of herbivores (%)	5	25	47	34	26	67	73	74	41	59
Mean individual body weight of all crustaceans (without *Leptodora*) (%)	n.d.	0.007	0.015	0.014	0.009	0.028	0.024	0.023	0.021	0.019
Mean biomass of *Leptodora* (mg w.w.)	0	0	0	0	0.003	0.017	0.045	0	0.138	0.014
Mean biomass of *Chaoborus* (mg w.w. · 1^{-1})	0	0	0	0	0	0	0.066	0.341	0.044	0.011

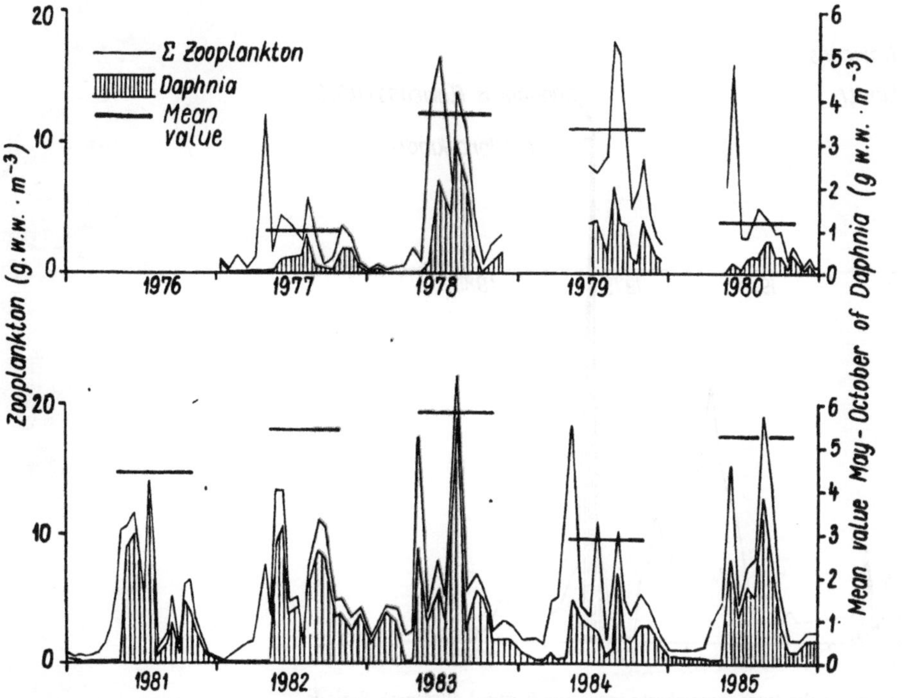

Fig. 4. Total biomass of the zooplankton and biomass of *Daphnia galeata* (vertical averages). Pre-biomanipulation period: 1976 to 1980, biomanipulation period: 1981 to 1985 (w.w. = wet weight). Notice the difference in the scales of the ordinates
(left: curves for total zooplankton and *D. galeata*; right: mean values of *D. galeata*).

(2) As Fig. 5 shows, even the mean individual body size of *Daphnia galeata* increased drastically during biomanipulation with the only exception of the second half of 1984 when a strong recruitment of perch occurred. It seems to be worth mentioning that just in such years, when great numbers of O+ perch occur the individual body size of *Daphnia galeata* is unusually low (compare curves of 1977 and 1984 in Fig. 5 with Fig. 2). These findings confirm the results from the Tjeukemeer in the Netherlands (Vijverberg and Van Densen 1984). Mainly as a consequence of the increased body size of *Daphnia*, the mean

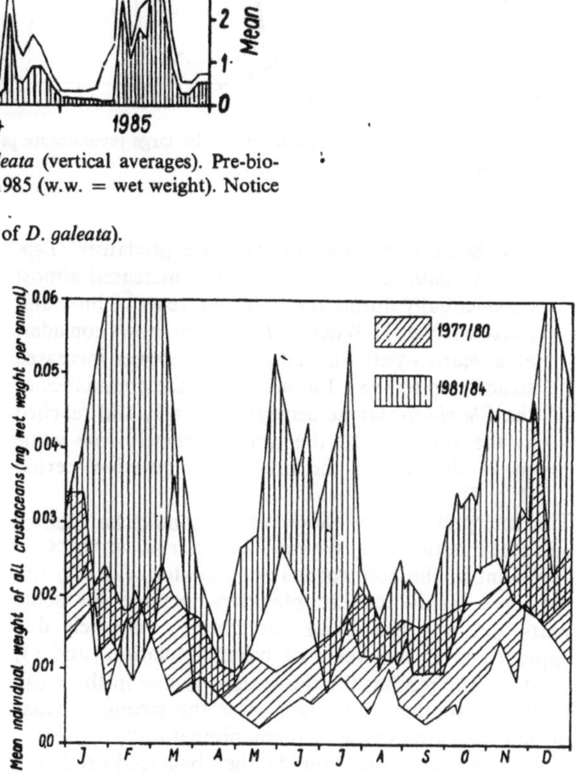

Fig. 6. Mean individual body weight of all crustaceans (including nauplii) Periods as in Fig. 4.

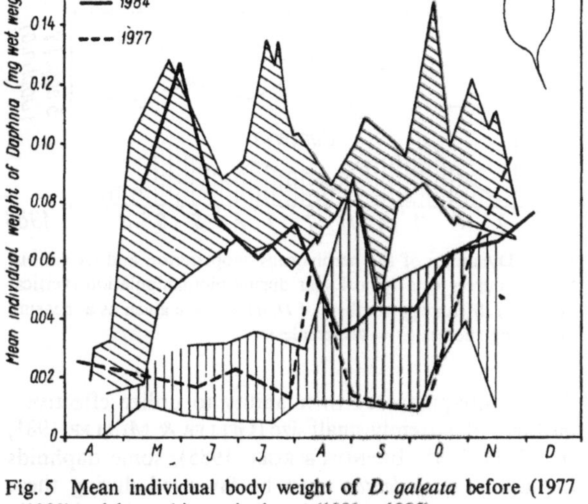

Fig. 5 Mean individual body weight of *D. galeata* before (1977 to 1980) and during biomanipulation (1981 to 1985).

individual body size of all crustaceans is also higher during biomanipulation compared with the pre-biomanipulation period (Fig. 6). The overlapping of the two bands in Fig. 6 is above all caused by high numbers of nauplii lowering sometimes considerably the mean body size of the crustaceans during biomanipulation. Other small-sized crustaceans contribute to this effect, expecially from the middle of July to the middle of September when *Chydorus sphaericus* and *Bosmina* species increase in number. But it cannot be neglected that the summer averages of the mean body size of all crustaceans steadily decreased during the biomanipulation period from 1981 to 1985 (see Table 3).

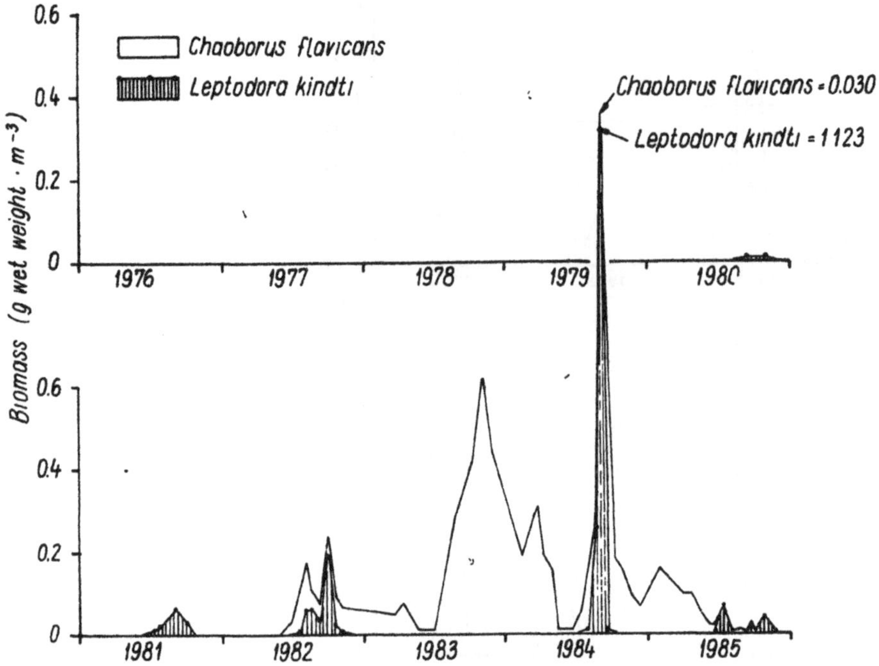

Fig 7. Development of the populations of the large invertebrate predators (vertical averages). Periods as in Fig. 4.

(3) The biomass of large invertebrate predators (*Leptodora kindti, Chaoborus flavicans*) increased almost exponentially during the first four years of biomanipulation (Fig. 7). Whereas *Leptodora* shows considerable year-to-year fluctuations, *Chaoborus* increased steadily up to 1983. But after 1983 the abundance of the *Chaoborus* larvae decreased sharply and reached at the end of 1985 the same extremely low value as at the end of the pre-biomanipulation period (Fig. 7).

The interpretation of these results should be focused on three questions: (1) Why did *Daphnia galeata* become dominant in the herbivorous zooplankton, (2) why did the large carnivorous zooplankters increase for some years and why has *Chaoborus* almost completely disappeared after five years of biomanipulation, and (3) what are the reasons for the steady decline in the mean body size of the crustaceans after the strong increase during the initial phase of biomanipulation?

With respect to question (1), the "balanced predation hypothesis" (Dodson 1974, Stenson 1976) could explain the shift from the dominance of small to large herbivores after 1980. But this hypothesis does not explain why just *Daphnia galeata* has become dominant (and not another large herbivore, e.g. *Eudiaptomus gracilis*). The explanation of this fact must be sought in the food situation in this hypertrophic reservoir. The high food concentration in spring (see Fig. 10) favours *Daphnia* due to their higher maximum feeding rates compared with *Eudiaptomus* (Lampert & Muck 1985). As a consequence of these high feeding rates of *Daphnia*, there are two periods during which algal food becomes very scarce: the clear water stage in June as well as late summer when only inedible algae occur in the phytoplankton (see Figs. 8 and 10). But even in these periods bacterial food and small-sized detritus particles are rather abundant (Fig. 8) as a result of the allochthonous organic load as well as zooplankton excretion and — during the late summer period — exudation of photosynthetic products by inedible algae.

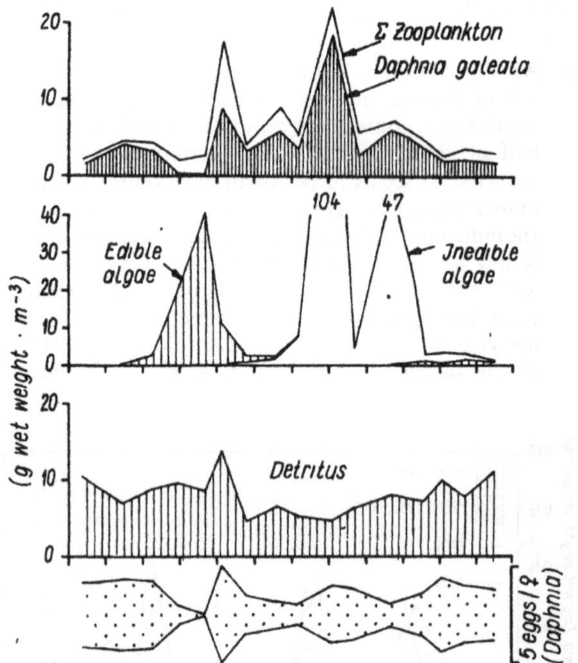

Fig. 8. Dynamics of the herbivorous zooplankton and its various food resources in a typical year during biomanipulation (vertical averages) Mean egg numbers of *D. galeata* are given as a measure of feeding success (♀ = all females).

As a consequence of their ability to utilize effectively particles of extremly small size (Geller & Müller 1981, de Mott 1985, Brendelberger 1985), some daphnids (including *D. galeata*) must be favoured during these periods. This assumption is confirmed by the relatively high egg numbers of *Daphnia galeata* (as a measure of feeding success), which are found in periods when bacterial food predominates (Fig. 8). Also in other investigations (PEDROS-ALIO & BROCK 1983, RIEMANN 1985), *D. galeata* was found to be among the most successful competitors under similar conditions. In addition to these aspects of the feeding mode, the life-history strategy of *D. galeata* (according to ROMANOVSKY 1984) seems to be best adapted to the food situation (including its variation in time) in the Bautzen reservoir.

Thus, the dominance of *Daphnia galeata* can only be interpreted as the result of the combined control of the zooplankton community structure by selective predation **and** food limitation (cf. STENSON 1984, BENNDORF & HORN 1985, GLIWICZ 1985, GOPHEN & POLLINGHER 1985). The great importance of the competition for food and its balance with predation also explains why the mean biomass of herbivores has not increased during the whole experiment (Table 3): exclusively the structure of the community has been changed by biomanipulation.

The first part of question (2) can easily be answered: Large invertebrate predators increase in number as a consequence of the lowered predation by fish. The same result has been found in other investigations (STENSON 1976, KERFOOT 1977, STENSON et al. 1978).

The second part of question (2) (disappearance of *Chaoborus* three years after its initial increase) is more complicated and shows the great significance of investigating long-term stability in food-web control. Most probably, the high amounts of young fish hatched in 1984 and 1985 have caused greater losses of the first instars of *Chaoborus* in these years compared with the first three years of the biomanipulation period. But, also in the case of the *Chaoborus* population (similar to *Daphnia*), the cooperative effects of predation and food limitation must be taken into consideration: As GILBERT (1985) has shown, rotifers cannot compete successfully with *Daphnia*, and their development is additionally inhibited due to mechanical injury as a consequence of the filtration process of *Daphnia* (GILBERT & STEMBERGER 1985). These findings are confirmed by the minimum abundances of all herbivorous rotifers in the Bautzen reservoir when the biomass of *Daphnia* is high in June (Fig. 9). Just at this time the first two instars of *Chaoborus* appear in the reservoir. Since these instars have a very high demand for food (KAJAK & RYBAK 1979) and rotifers are a main component of its food, it is most likely that high mortality of the young instars is caused by starvation. NEILL (1985) was able to show experimentally this indirect mechanism of exclusion of *Chaoborus* by *Daphnia*. But this exclusion can only be expected in oligotrophic lakes (as in the experimental water of NEILL 1985) and in slightly stratified or unstratified lakes of higher trophic states, i.e. in water bodies without oxygen depletion in the deep layers. In water bodies having a hypolimnion free of oxygen this mechanism of indirect exclusion of *Chaoborus* by *Daphnia* cannot occur due to spatial separation of *Daphnia* from some small grazers (some rotifers, large protozoans) which tolerate extremly low oxygen concentrations near the thermocline. Thus, even at a high abundance of *Daphnia* enough small-sized zooplankton is available as food for the smallest instars of *Chaoborus* in water bodies with a stable oxygen stratification (e.g. BENNDORF et al. 1984, BÖTTGER & KÖHLER 1986). Taking into consideration the fact that Bautzen reservoir was fairly stably stratified in 1982 and 1983, but almost unstratified in 1984 and 1985 (due to unusually cool and windy weather as well as an extremely low water level), it becomes evident that the reason for the up-and-down development of the *Chaoborus* population is rather complex: it must be understood as the combined effect of changing predation by fish and year-

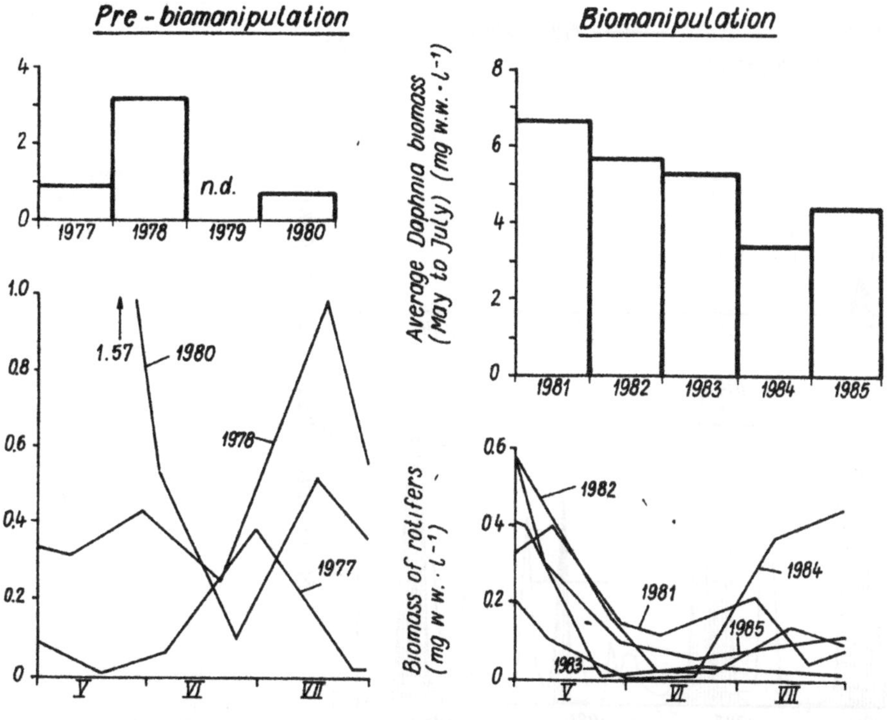

Fig. 9. Comparison of the biomass of *D. galeata* (mean values from May to July) with the biomass of rotifers during the same period (vertical averages): competitive exclusion of rotifers by *D. galeata* during the clear water stage in June under conditions of biomanipulation (see text for explanation). (n.d. = not determined; w.w. = wet weight).

to-year variations of indirect exclusion of rotifers (and nauplii) by *Daphnia*, both the factors being controlled by the meterological and hydrological constellation. Thus, a long-term stability in the *Chaoborus* population cannot be expected in Bautzen reservoir (and similar water bodies which are subjected to strong year-to-year changes in their physical structure).

There is no long-term trend in the abundance of the *Leptodora* population during biomanipulation. With the exception of a single sampling date in 1984 (see Fig. 7), *Leptodora* is effectively controlled by fish (mainly Percidae, cf. KÖPKE et al. 1988), thus revealing that even a reduced stock of zooplankton-eating fish during biomanipulation is able to keep in check large invertebrate predators with short generation times. There is no sufficient explanation for the obvious exclusion of *Leptodora* when the maximum of *Chaoborus flavicans* occurs in 1983 (Fig. 7).

As a consequence of the instability of the abundance of the invertebrate predators and the year-to-year variations in fish recruitment, the mean individual body size of the crustaceans could not be stabilized at the desired high level. But even the lowest value of the biomanipulation period (0.019 mg w.w in 1985) is higher than the maximum value during the pre-biomanipulation period (0.015 mg w.w. in 1978; Table 3). At this stage of the experiment it remains unclear whether the steady decline in the mean individual body size of the crustaceans from 1981 to 1985 (question 3) represents a long-term trend or an accidental coincidence of unfavourable circumstances in 1984 and 1985. The latter explanation is rather probable because the sharp decline of piscivorous perch in 1984 and the higher intensity of vertical mixing in 1984 and 1985 are independent processes which, nevertheless, have the same effects: increasing fish predation and decreasing *Chaoborus* predation on herbivorous zooplankton, both processes causing a lower individual body size of the herbivores.

Effects on phytoplankton and water quality

Immediatly after the melting of the ice cover, diatoms (*Asterionella formosa*, *Stephanodiscus hantzschii*, *Cyclotella meneghiniana*) usually reach high biomasses between 20 and 40 mg wet weight · 1^{-1}. With the exception of the year 1984 (unfavourable meterological conditions), this spring maximum of the diatoms occurs in all the years of the investigation without any significant difference between the pre-biomanipulation and the biomanipulation period (Fig. 10). During late April and the beginning of May this diatom bloom is drastically reduced by zooplankton grazing, mainly by *Cyclops vicinus*, rotifers, and in May additionally by cladocerans.

After the spring maximum of the diatoms, the phytoplankton development differs strongly from year to year (Fig. 11). In **pre-biomanipulation** years the spring maximum is immediately followed by a rather strong development of Chlorococcales, Cryptomonadales and diatoms, or blue greens, respectively. There are no clear water stages in these years (see Fig. 13: 1976 to 1980). During the **biomanipulation** period, in all years (with the exception of 1985) clear water stages are observed after the diatom bloom. They are closely related to the development of the *Daphnia* population and characterized by extremly low phytoplankton biomasses and maximum Secchi depths between 4 und 7 m. Depending on the zooplankton (*Daphnia*) biomass during these clear water stages, the oxygen concentration (at noon) is lowered to about 70% to 80% saturation. Similar situations have been reported by MOTHES & KROCKER (1986) from a lake which is comparable with the Bautzen reservoir.

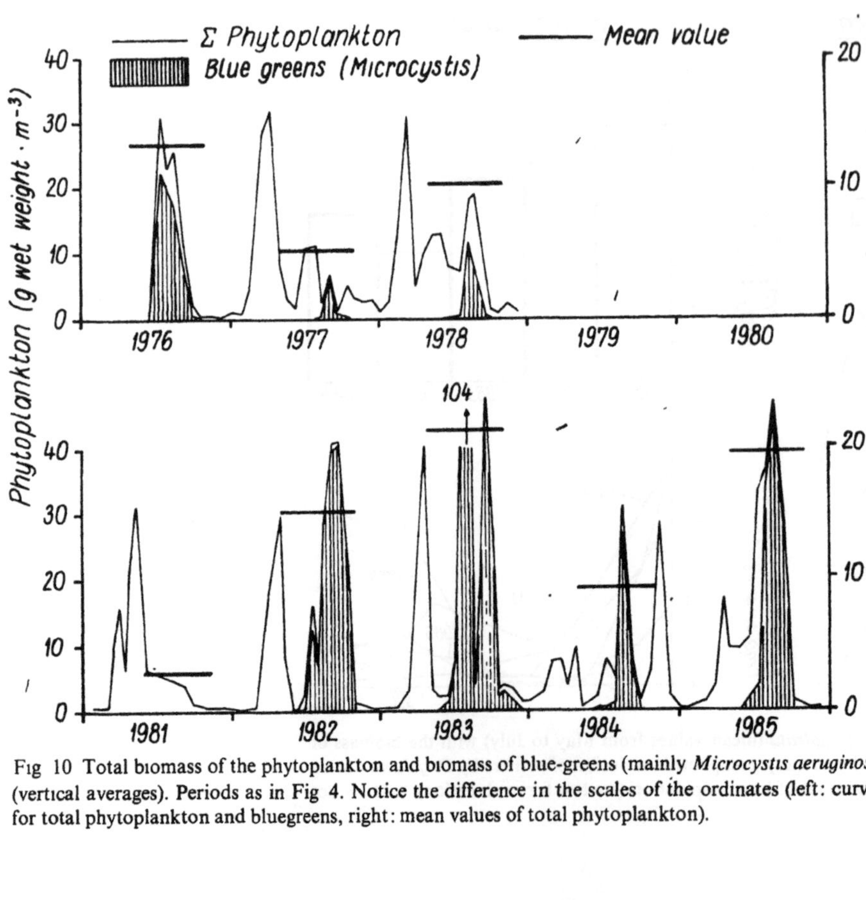

Fig 10 Total biomass of the phytoplankton and biomass of blue-greens (mainly *Microcystis aeruginosa*) (vertical averages). Periods as in Fig 4. Notice the difference in the scales of the ordinates (left: curves for total phytoplankton and bluegreens, right: mean values of total phytoplankton).

Extremely high oxygen consumption rates of the *Daphnia* population, which are well-known from carp ponds (Barthelmes 1981), are avoided in Bautzen reservoir even during biomanipulation due to the remaining moderate fish predation on zooplankton. The clear water stage persists till late June or the beginning of July (see Fig. 8: phytoplankton).

From July onwards, the phytoplankton community can develop in two completely different ways (Fig. 11), which seem to be controlled mainly by meterological conditions: In years with cool and windy summer conditions, large, inedible green algae dominate as far as the biomanipulation period is concerned (1981: *Hydrodictyon*; 1984, 1985: *Pediastrum*). In 1981 these algal mats were quickly removed by grazing of grass carps (*Ctenopharyngodon idella*) and a new clear water stage persisted till late autumn (Fig. 13). But in 1984 and 1985, when also large green algae had dominated

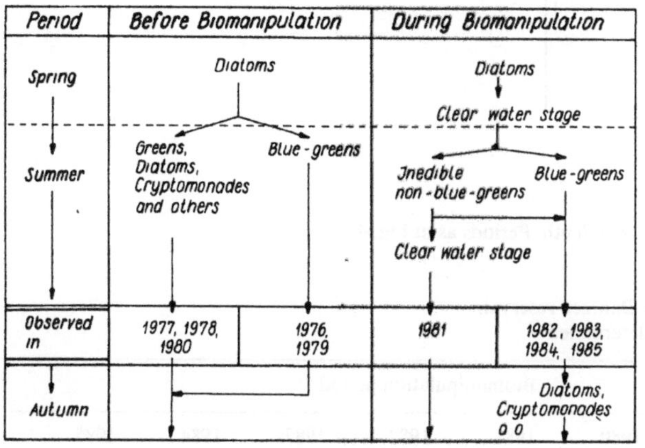

Fig. 11. Types of phytoplankton successions in the Bautzen reservoir before and during biomanipulation.

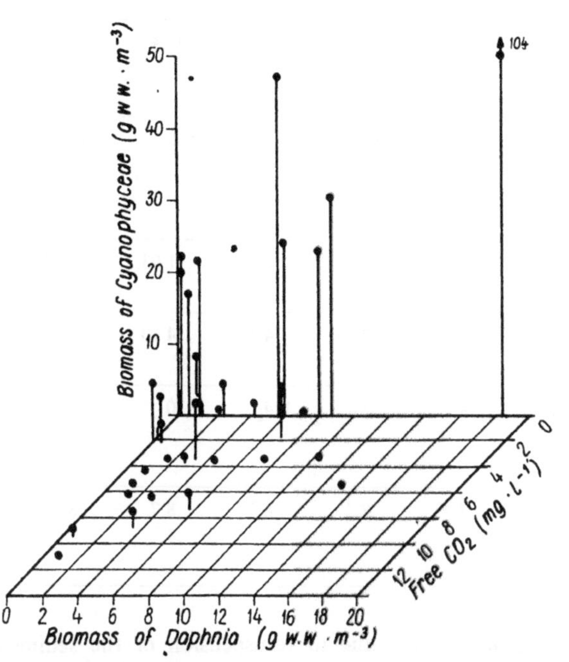

Fig. 12. Dependence of the biomass of blue-greens (mainly *Microcystis aeruginosa*) on the grazing pressure exerted by *D. galeata* and on the content of the free CO_2. Only periods with potential mass developments of blue-greens (June to September) are considered (Free CO_2: concentration at the surface; biomasses: vertical averages).

during the first half of the summer, they were replaced by *Microcystis* in August obviously due to a change in the meterological situation (low wind speed, high solar radiation). An assemblage of various edible algae develops during summer in pre-biomanipulation years if the meteorological situation is characterized by cool and windy weather (1977, 1978, 1980).

If, on the other hand, summer conditions are dominated by high solar radiation and long periods of low intensity of the vertical mixing processes, *Microcystis aeruginosa* already becomes dominant during early summer in years before as well as during biomanipulation (1976, 1979, 1982, 1983). Fig. 12 demonstrates that *Microcystis* dominance is primarily dependent on the content of free CO_2 in the euphotic zone rather than on the grazing pressure exerted by *Daphnia*. A sudden increase in the vertical mixing intensity due to a storm, and the resulting introduction of free CO_2 from the deep layers to the surface, has caused a fast collapse of the *Microcystis* blooms in all the cases observed in Bautzen reservoir (Benndorf & Herrmann, in prep.). There are some indications that a disregulation of the photosynthesis/respiration relationship occurs in *Microcystis* under these conditions, but direct and/or indirect effects of the pH-decrease due to the increase in free CO_2 must also be taken into consideration (Shapiro 1984).

It is not possible to evaluate the consequences of biomanipulation with respect to water quality in the Bautzen reservoir without taking into account the extremely high external phosphorus load (Fig. 14) which enables the phytoplankton to grow without any nutrient limitation throughout the whole year. Under these conditions, where there is a continuously high nutrient import, no decrease of the in-lake phosphorus concentration can be expected. Even if one assumes increased P-losses during biomanipulation due to vertical migration of *Daphnia* (as shown by Wright & Shapiro 1984), this effect is superimposed — in addition to the high P-import — by the reduced phosphorus uptake by the algal community due to (1) extended clear water stages, and (2) — during surface blooms of *Microcystis* — a drastically reduced depth of the layer of high primary production. The reason of this latter effect is the almost complete absence of algae below the surface blooms ("cryptic" clear water stage according to Barthelmes 1981, Mothes & Krocker 1986). As Fig. 14 and Table 4 show, the reduced P-uptake must play the more important role when compared with the simultaneously increased P-losses: The vertical mean in-lake concentration of dissolved orthophosphate has increased as an average during biomanipulation by a factor of 2.9 compared with the pre-biomanipulation period. Since the external load has also increased, the more interesting "net" orthophosphate increase, which is exclusively caused by in-lake processes, must be calculated from the following ratio: in-lake concentration/external load. Even this calculation results in a factor of the "net" increase of the in-lake PO_4-P concentration of 1.7, thus demonstrating the lower phosphate elimination in the Bautzen reservoir as a consequence of biomanipulation. Rather similar results have been found with respect to total phosphorus ("net" increase due to biomanipulation by a factor of 1.5).

The summer averages as well as the maximum values of the total phytoplankton biomass during the biomanipulation period depend on the type of phytoplankton succession (cf. Fig. 11) and they are either lower (1981), equal (1984) or higher (1982, 1983, 1985) compared

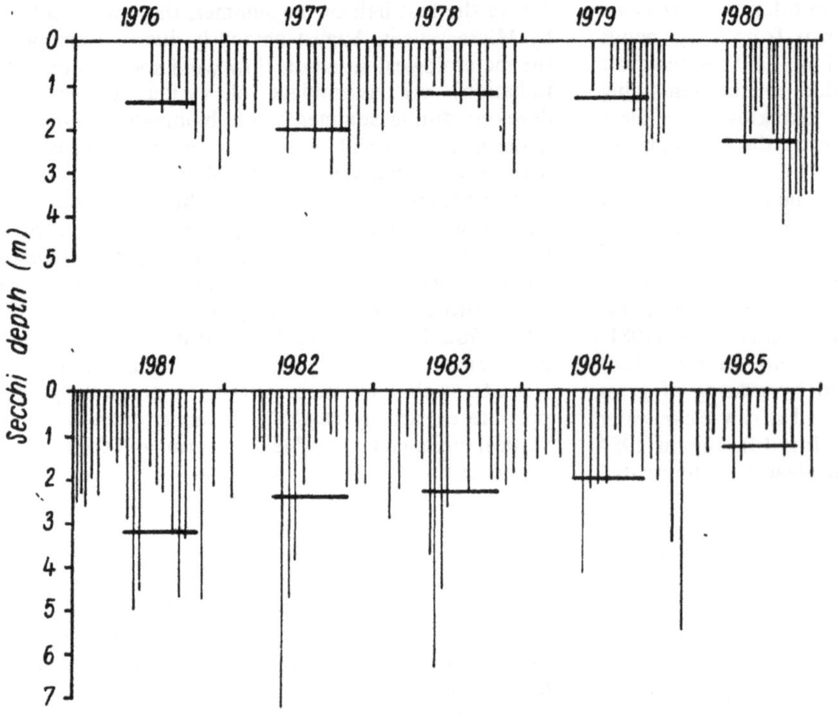

Fig. 13. Single values and summer averages (May to October) of the Secchi depth Periods as in Fig. 4.

Table 4. Characteristics of the phytoplankton and water quality in the Bautzen reservoir [values from May to October; *n* (per season) = 9 ... 12], n.d. = not determined

Criteria	Pre-biomanipulation period					Biomanipulation period				
	1976	1977	1978	1979	1980	1981	1982	1983	1984	1985
Mean phytoplankton biomass (mg w.w. $\cdot l^{-1}$)	13.3	5.3	10.2	n.d.	n.d.	3.0	17.4	21.6	9.2	19.7
Maximum phytoplankton biomass (mg w.w. $\cdot l^{-1}$)	30.5	11.2	18.5	n.d.	n.d.	6.5	40.9	104.6	30.4	47.3
Mean in-lake concentration of orthophosphate (mg P $\cdot m^{-3}$)	219	109	83	n.d.	n.d.	n.d.	n.d.	366	560	278
Mean Secchi depth (m)	1.41	2.00	1.30	1.00	2.33	3.02	2.42	2.26	1.93	1.22
Maximum Secchi depth (m)	2.20	3.00	1.80	2.50	4.20	5.00	7.20	4.50	4.10	2.00
Mean O_2-saturation at the surface, 12°° a.m. (%)	97	107	148	155	n.d.	119	135	114	120	103
Mean pH at the surface, 12°° a.m.	8.8	8.1	9.0	9.2	n.d.	7.4	8.2	8.3	8.7	8.3

with the equivalent values of the pre-biomanipulation period (Table 4 and Fig. 10). Nevertheless, the Secchi depths have been increased by biomanipulation in almost all years as a consequence of the dominance of large algae colonies (Table 4 and Fig. 13). The relatively bad situation in 1985 was caused by the extremly low water-level during that year which led to an unusually high detritus content due to resuspension of the sediment and an increased primary production per unit of the whole water column. As the summer averages of the surface pH and oxygen saturation show (Table 4), a slight improvement in these water quality criteria during biomanipulation can also be stated, mainly as a consequence of the clear water stages.

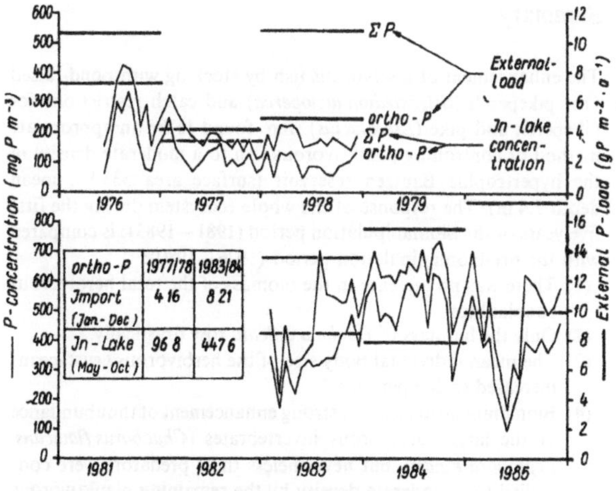

Fig. 14. External areal load and in-lake concentration of phosphorus (ΣP = total phosphorus, ortho-P = dissolved reactive phosphorus; method: molybdate-sulfuric acid and ascorbic acid) No determinations from 1979 to 1982. Table: Comparison of two typical years before (1977, 1978) and during (1983/84) biomanipulation. Periods as in Fig. 4.

Discussion

As EDMONDSON & ABELLA (1988) have emphasized, ways should be developed to change fish populations by methods less drastic than poisoning the whole lake with rotenone. The intentional enhancement of the native populations of piscivorous fish (*Stizostedion lucioperca* and *Esox lucius*) in the Bautzen reservoir is such a less drastic method which was found to be appropriate to control planktivorous fish to a moderate density. The increased predation pressure of these large piscivores supports the development and maintenance of a moderate abundance of the perch population (*Perca fluviatilis*) with a sufficiently high proportion of piscivorous age-classes. This seems to be a key factor in controlling the young-of-the-year of all fish (including its own species).

The remaining stock of planktivorous fish prevents an excessive development of large carnivorous invertebrates. As a consequence of the lowered predation by fish and of the simultaneously controlled predation by large carnivorous invertebrates, there is a high probability that the biomass of large herbivores (mainly *Daphnia*) is increased by this kind of food-web manipulation (Fig. 15A). The combined control of selective predation by fish and carnivorous invertebrates on the one hand and limitation of the available food for herbivores during summer on the other provides an explanation for the dominance of *Daphnia*.

Whereas *Leptodora kindti* is able to produce short peaks in abundance when the predation by fish temporarily decreases as the consequence of extrem pH-values (see Fig. 7: 1984), the abundance of *Chaoborus* larvae increases over the first three years of biomanipulation and decreases during the fourth and fifth year to very low numbers. This decrease of *Chaoborus flavicans* is the consequence of the predation by 0+ perch and 0 + roach as well as of the starvation of the first and second instars of *Chaoborus*. Starvation is caused by the competitive exclusion of rotifers and nauplii by *Daphnia*. Thus, the desirable moderate predation pressure of *Chaoborus* on small- and medium-sized herbivores is inhibited automatically by the dominance of *Daphnia*. This negative feedback mechanism indicates that only an oscillating long-term stability of biomanipulation can be expected as far as the mean body size of herbivores is concerned.

The increase in number and body size of *Daphnia* does not reach such an undesired extent that it could cause oxygen depletion as described from carp ponds (e.g. BARTHELMES 1981). It must be emphasized that this possible negative side-effect of biomanipulation under hypertrophic conditions can be avoided only if an excessive development of *Daphnia* is controlled by the moderate density of planktivorous fish. Thus, for several reasons (prevention of large invertebrate predators, starvation of *Daphnia*, oxygen depletion), the stability of the desirable structure of the zooplankton community can be guaranteed by a moderate (low) stock of planktivorous fish rather than by an extremination of these fish (e.g. by poisoning).

Despite the considerable enhancement of the filtration capacity of the zooplankton, there is no stable reduction of the phytoplankton biomass during the biomanipulation period. Whereas before biomanipulation those algae dominated, which are characterized by high maximum growth rates, during biomanipulation algae become dominant the maximum growth rates of which are rather low. High biomass of this latter type of algae is reached by minimizing their grazing losses due to effective mechanisms of protection against feeding by zooplankton. These findings confirm the results of many other investigations (MC CAULEY & BRIAND 1979, LYNCH & SHAPIRO 1981, LAMPERT 1981, ANDERSSON & CRONBERG 1984, BENNDORF et al. 1984).

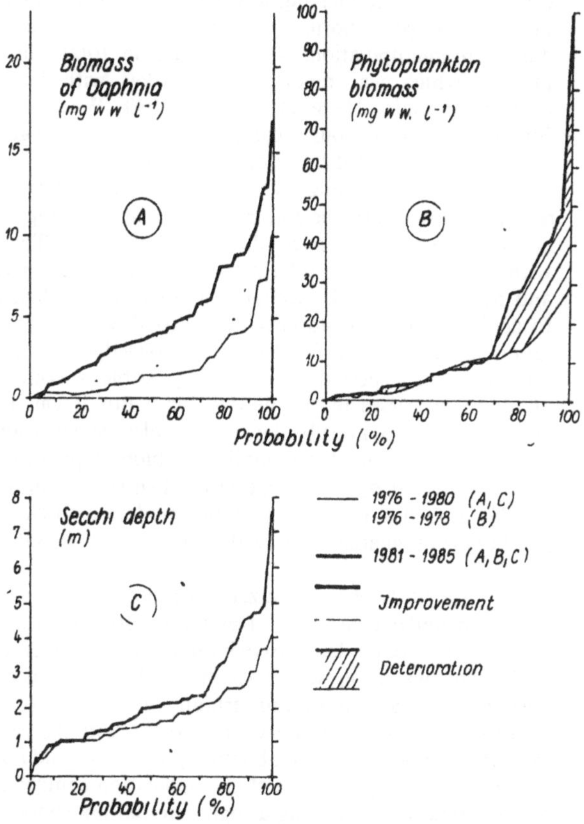

Fig 15 Comparison of the probability distribution (duration curves) of three important criteria before (1976 to 1980) and during (1981 to 1985) biomanipulation. Abscissa: probability that a certain value of the ordinate is not exceeded.

Under hypertrophic conditions and if the meteorological situation favours the development of *Microcystis*, which is very often extremely well protected against grazing (colony size, toxicity), the probability of extremely high phytoplankton biomasses is increased by biomanipulation. This becomes evident from Fig. 15B. It must be stressed that this conclusion is not a contradiction of the statement concerning the primary control function of the vertical mixing processes and the role of free CO_2 (and pH) with respect to the dominance of *Microcystis* in a phosphorus-rich environment (cf. Fig. 12). These physical and chemical conditions are the primary prerequisites, but if they are fulfilled, strong zooplankton grazing seems to enhance the probability of an extreme *Microcystis* bloom (1982, 1983, 1985). If the physical and chemical prerequisites are not fulfilled (as in 1981 and 1984), such extreme blooms cannot occur, and the results of 1981 demonstrate clearly that under favourable circumstances biomanipulation can cause considerable improvement in the water quality throughout the whole year even in a hypertrophic water body having a continuously high external nutrient load. But this desirable result has only been found in one out of five years.

Despite of the increased probability of extremely high phytoplankton biomasses, the transparency of the water has been improved during 75 per cent of the time of the biomanipulation period compared with the years before biomanipulation (Fig. 15C). But it cannot be overlooked that just during the months with the most intensive recreational use of the reservoir (July and August) the probability of extreme *Microcystis* blooms is higher than the probability of increased transparency.

As a consequence of extended clear water stages and the dominance of algae having low rates of growth and P-uptake, the phosphorus concentration has increased during biomanipulation. The decrease in total phosphorus, which was reported from lakes with lower external P-loads (HENRIKSON et al. 1980, WRIGHT & SHAPIRO 1984), obviously cannot be expected in water bodies heavily loaded by external P-sources.

The results of the long-term experiment in the Bautzen reservoir show clearly the dominance of indirect ecological control mechanisms over direct mechanisms. Thus, without taking into consideration the numerous positive and negative feedback mechanisms, there is no chance to achieve long-term stability of the food-web manipulation. At the present level of our knowledge of these indirect control mechanisms, it is impossible to control all these processes in the desired way. But it can be concluded from our results that a maximal approximation to this aim (i.e. long-term stability of biomanipulation) can be achieved by manipulating the fish stock continuously in such a way that a stable moderate stock of planktivores is guaranteed (see also EDMONDSON & ABELLA 1988).

The case study of the Bautzen reservoir demonstrates that even under conditions of extreme nutrient saturation the principal mechanisms of biomanipulation operate in the same way as in waters which are characterized by lower trophic states. But the uncertainty of the response of the ecosystem to biomanipulation increases with increasing nutrient load. Therefore, biomanipulation should not be used as the only measure of eutrophication control in hypertrophic waters having a high external phosphorus load. A combination of this measure with other control possibilities (e.g. destratification, reduction of the phosphorus load) seems to be inevitable in these water bodies.

Summary

The enhancement of piscivorous fish by stocking with pond-raised 0+ pikeperch (*Stizostedion lucioperca*) and catch restrictions for pikeperch and pike (*Esox lucius*) were found to be an appropriate method of controlling planktivorous fish to a moderate density in the hypertrophic Bautzen reservoir (surface area 533 ha, mean depth 7.4 m). The response of the whole ecosystem during the first five years of the biomanipulation period (1981—1985): is compared with the pre-biomanipulation period (1976—1980):

(1) There was no increase in the biomass of the total herbivorous zooplankton.
(2) Only the biomass of *Daphnia galeata* rose drastically.
(3) The mean individual body size of the herbivorous crustaceans increased to 209 per cent.
(4) Biomanipulation led to a strong enhancement of the abundance of the large carnivorous invertebrates (*Chaoborus flavicans*, *Leptodora kindti*), but nevertheless these predators were controlled to a moderate density by the remaining planktivorous fish
(5) Extended clear water stages with maximum Secchi depth of 4 to 7 m were observed in early summer during the biomanipulation period; no clear water stages occurred before 1980
(6) The mean Secchi depth (May to October, 1981 to 1985) increased by 0 56 m (i.e. 35%).
(7) Inedible algae became dominant during summer of all the years of biomanipulation, and the mean as well as the maximum phytoplankton biomasses are either lower (1981), equal (1984), or higher (1982, 1983, 1985) than during the pre-biomanipulation period
(8) The probability of the occurrence of extremely dense *Microcystis* blooms has been raised by the increased grazing pressure of *Daphnia* on the phytoplankton.
(9) Total phosphorus increased to 150 per cent as an average
(10) Some responses of the ecosystem show remarkable instability during the biomanipulation period (e g. steady decrease of the mean individual body size of the herbivorous crustaceans, strong up- and downward trends in the abundance of *Chaoborus* and strong fluctuations in the abundance of *Leptodora*, bistability of the structure of the summer phytoplankton). These instabilities are regarded to be the consequence of the stochastic character of the meteorological conditions as well as the result of the time lag of the feedback mechanisms within the ecosystem.

The conclusion is drawn that the principal mechanisms of biomanipulation in hypertrophic waters heavily loaded by external phosphorus sources are almost the same as in waters having lower trophic states But the probability of an improvement of the water quality by biomanipulation decreases with increasing P-load due to the high number of "degrees of freedom" of a nutrient-saturated ecosystem. Therefore, effective eutrophication control of these water bodies seems to be possible only by combining food-web manipulation with other control measures (e.g reduction of the external P-load, destratification). The maximum possible long-term stability of food-web manipulation can be achieved by controlling the fish community in such a way that a stable moderate stock of planktivorous fish is guaranteed rather than by exterminating these fish (e.g. by poisoning)

Zusammenfassung

Es werden 5-jährige Erfahrungen (1981—1985) mit der Biomanipulation mit Hilfe des Einsatzes von Zandern und Fangeinschränkungen für Zander und Hecht in der hypertrophen Talsperre Bautzen mitgeteilt. Die Wassergüte vor und nach der Biomanipulation wird verglichen. Die Kontrolle der planktonfressenden Fische durch Raubfische ist ihrer totalen Beseitigung vorzuziehen. Sie ist aber bei der extrem hohen Nährstoffbelastung noch durch andere Methoden zu ergänzen (z. B. Reduktion der P-Last, Destratifikation)

Acknowlegements

The authors wish to thank Dr. K JESCHKE, Dipl.-Ing. K. PÜTZ, Dipl.-Ing. D GLOS, Dipl.-Fischw. W. SARODNIK, W. LOCHMANN, Dr J. HOFMANN, Dipl -Fischw. N. LANGNER and Dr. I. FRITZSCHE for their valuable assistance and cooperation in organizing the field experiment. G EGERER, G. HERRMANN, E. BALL, M. GEBHART and S. FUCHS are gratefully acknowledged for technical assistance. We thank BA D COCKELL for improving the English text.

References

ANDERSSON G & CRONBERG, G. (1984): *Aphanizomenon flos-aquae* and large *Daphnia* — An interesting plankton association in hypertrophic waters In BOSHEIM, S., & NICHOLLS, M. (eds.), Interaksjoner mellom trofiske nivaer i ferskvann, Norsk Limnologforening, Oslo, 63—76.

BARTHELMES, D. (1981): Hydrobiologische Grundlagen der Binnenfischerei. Jena.

BENNDORF, J. (1979). Kausalanalyse, theoretische Synthese und Simulation des Eutrophierungsprozesses in stehenden und gestauten Gewässern. Diss. B, Techn. Univ. Dresden, Fakultät f. Bau-, Wasser- u. Forstwesen, 1—165.

— (1988). Preface to the issue "Biomanipulation". *Limnologica* (Berlin) **19** (1), 5—8.

— & HERRMANN, G. (in prep.): Control of mass developments of blue-green algae by carbon dioxide in Bautzen reservoir (GDR). Int Revue ges. Hydrobiol. (in prep).

— & HORN, W. (1985): Theoretical considerations on the relative importance of food limitation and predation in structuring zooplankton communities. Arch. Hydrobiol. Beih. Ergebn. Limnol. **21**, 383—396.

— KNESCHKE, H., KOSSATZ, K., & PENZ, E. (1984). Manipulation of the pelagic food web by stocking with predacious fishes. Int. Revue ges. Hydrobiol. **69**, 407—428.

BÖTTGER, A., & KÖHLER, J. (1986). Langzeiteffekte der Biomanipulation im Steinbruch-Restgewässer Gräfenhain. Jahresarb. Techn. Univ. Dresden, Sektion Wasserwesen, 1—31.

BOTTRELL, H. H., DUNCAN, A., GLIWICZ, Z. M., GRYGIEREK, E., HERZIG, A., HILLBRICHT-ILKOWSKA, A.; KURASAWA, H., LARSSON, P., & WEGLENSKA, T. (1976): A review of some problems in zooplankton production studies. Norw. J. Zool. **24**, 419—456.

BRENDELBERGER, H. (1985): Filter mesh-size and retention efficiency for small particles: comparative studies with Cladocera. Arch. Hydrobiol., Beih. Ergebn. Limnol. **21**, 135—146

DEMOTT, W. R. (1985): Relations between filter mesh-size, feeding mode, and capture efficiency for cladocerans feeding on ultrafine particles. Arch. Hydrobiol., Beih. Ergebn. Limnol. **21**, 125—134.

DODSON, S. C. (1974): Zooplankton competition and predation: an experimental test of the size-efficiency hypothesis Ecology **55**, 605—613.

EDMONDSON, W. T., & ABELLA, S. (1988): Unplanned biomanipulation in Lake Washington *Limnologica* (Berlin) **19** (1), 73—79.

GELLER, W., & MÜLLER, H. (1981): The filtration apparatus of Cladocera: filter mesh-sizes and their implications on food selectivity. Oecologia (Berlin) **49**, 316—321

GILBERT, J. J. (1985): Competition between rotifers and *Daphnia*. Ecology **66**, 1943—1950.

— & STEMBERGER, R. S. (1985): Control of *Keratella* populations by interference competition from *Daphnia*. Limnol. Oceanogr. **30**, 180—188.

GLIWICZ, Z. M. (1985): Predation or food limitation: an ultimate reason for extinction of planktonic cladoceran species. Arch. Hydrobiol., Beih. Ergebn. Limnol. **21**, 419—430.

GOPHEN, M., & POLLINGHER, U. (1985): Relationships between food availability, fish predation and the abundance of the herbivorous zooplankton community in Lake Kinneret. Arch. Hydrobiol., Beih. Ergebn. Limnol. **21**, 397—405.

HENRIKSON, L., NYMAN, H. G., OSCARSON, H. G., & STENSON, J. A. (1980): Trophic changes, without changes in the external nutrient loading. Hydrobiologia **68**, 257—263.

HORN, W. (1978): Der Anteil des Crustaceenplanktons an der Verringerung des Phytoplanktongehalts in der Saidenbachtalsperre. Diss. A, Techn Univ. Dresden, Fakultät f. Bau-, Wasser- u. Forstwesen, 1—163.

HORNIG, U., & BENNDORF, J. (1985): Biomanipulation in the Bautzen reservoir: spring zooplankton succession and consequences of spatial inhomogeneity in fish predation on summer zooplankton J. Hydrobiol. **1**, 53—61

HRBAČECK, J., DVORAKOVA, M., KORCINEK, V., & PROCHAZKOVA, L. (1961): Demonstration of the effect of fish stock on the species composition and the intensity of metabolism of the whole plankton association. Verh. Int. Verein. Limnol. **14**, 192—195.

KAJAK, Z., & RYBAK, J. (1979): The feeding of *Chaoborus flavicans* MEIGEN (Diptera, Chaoboridae) and its predation on lake zooplankton. Int. Rev. ges Hydrobiol. **64**, 361—378.

KERFOOT, W CH (1977): Implications of copepod predation. Limnol Oceanogr **22**, 316—325.

KÖPKE, U., SCHULTZ, H , JARCHOW, R., HORNIG, U., & PENIG, J. (1988) Analyse des Nahrungskonsums von Barschen (*Perca fluviatilis*) in der Talsperre Bautzen *Limnologica* (Berlin) **19** (1), 37—43.

LAMPERT, W. (1981): Inhibitory and toxic effects of blue-green algae on *Daphnia*. Int. Rev. ges. Hydrobiol. **66**, 285—298.

— & MUCK, P. (1985): Multiple aspects of food limitation in zooplankton communities: The *Daphnia-Eudiaptomus* example. Arch. Hydrobiol., Beih. Ergebn. Limnol. **21**, 311—322.

LESSMARK, O (1983): Competition between perch (*Perca fluviatilis*) and roach (*Rutilis rutilus*) in south Swedish lakes. Doct. Diss., Lund Univ , Inst. of Limnology, 1—172

LYNCH, M., & SHAPIRO, J. (1981): Predation, enrichment, and phytoplankton community structure. Limnol Oceanogr. **26**, 86—102.

MCCAULEY, E., & BRIAND, F. (1979). Zooplankton grazing and phytoplankton species richness: Field tests of the predation hypothesis. Limnol. Oceanogr. **24**, 243—252.

MOTHES, G., & KROCKER, M. (1986): Beziehungen zwischen der Wasserqualität des Großen Müggelsees und der Zooplanktonentwicklung Acta hydrochim. hydrobiol. **14**, 263—282.

NEILL, W. E. (1985): The effects of herbivore competition upon the dynamics of *Chaoborus* predation. Arch. Hydrobiol., Beih. Ergebn. Limnol. **21**, 483—491.

PEDROS-ALIO, C., & BROCK, T. D. (1983): The impact of zooplankton feeding on the epilimnetic bacteria of an eutrophic lake. Freshwater Biology **13**, 227—239.

PENIG, J. (1983): Untersuchungen zum Nahrungskonsum der Hauptfischarten in der Talsperre Bautzen und Saidenbach. Dipl.-Arb., Techn. Univ. Dresden, Sektion Wasserwesen, 1—75.

PERSSON, L. (1983a): Effects of intra- and interspecific competition on dynamics and size structure of a perch *Perca fluviatilis* and a roach *Rutilus rutilus* population. Oikos **41**, 126—132.

— (1983b): Food consumption and the significance of detritus and algae to intraspecific competition in roach *Rutilus rutilus* in a shallow eutrophic lake. Oikos **41**, 118—125.

RECKNAGEL, F. (1984): Angewandte Systemanalyse in der Wassergütebewirtschaftung von Standgewässern. Diss. B, Techn. Univ. Dresden, Fakultät f. Bau-, Wasser- und Forstwesen, 1—169.

RIEMANN, B. (1985): Potential importance of fish predation and zooplankton grazing in natural populations of freshwater bacteria. Appl. Environm Microbiol. **50**, 187—193

ROMANOVSKY, Y. E. (1984): Individual growth rate as a measure of competitive advantages in cladoceran Crustaceans. Int. Rev. ges. Hydrobiol. **69**, 613—632.

SCHULTZ, H. (1983): Untersuchungen über die Größe und Struktur des Fischbestandes sowie den Nahrungskonsum ausgewählter Fischarten in zwei Stauseen unterschiedlichen Trophiegrades. Diss. A, Techn. Univ Dresden, Fakultät für Bau-, Wasser- u. Forstwesen, 1—141.

— (1987): An acoustic fish stock assessment in the Bautzen reservoir. *Limnologica* (Berlin) **19** (1), 61—70.

SHAPIRO, J. (1984): Blue-green dominance in lakes: The role and management significance of pH and CO_2. Int Revue ges. Hydrobiol. **69**, 765—780.

— LAMARRA, V., & LYNCH, M. (1975): Biomanipulation · An ecosystem approach to lake restoration. In: BREZONIK, P. L., & FOX, J. L. (eds.), Water Quality Management through Biological Control. Report No. ENV-07-75-1, University of Florida, Gainesville, 85—96.

STENSON, J. A. E. (1976): Significance of predator influence on composition of *Bosmina* spp. populations. Limnol. Oceanogr. **21**, 814—822.

—: Interactions between pelagic metazoan and protozoan zooplankton, an experimental study. Hydrobiologia **111**, 107—112.

— BOHLIN, T., HENRIKSON, L., NILSSON, B. I., NYMAN, H. G., OSCARSON, H. G., & LARSSON, P. (1978): Effects of fish removal from a small lake. Verh. Int. Verein. Limnol. **20**, 794—801.

UTERMOHL, H. (1958): Zur Vervollkommnung der quantitativen Phytoplankton-Methodik. Mitt. Int. Verein. Limnol. **9**, 1—38.

VIJVERBERG, J, & DENSEN, W L T VAN (1984): The role of the fish in the food web of Tjeukemeer, The Netherlands. Verh Int. Verein. Limnol. **22**, 891—896.

WRIGHT, D. I., & SHAPIRO, J. (1984): Nutrient reduction by biomanipulation: An unexpected phenomenon, and its possible cause. Verh. Int. Verein. Limnol. **22**, 518—524.

ZARET, T. M. (1980): Predation and freshwater communities. Yale Univ. Press, New Haven, Conn.

Received December 3, 1986

Address of the responsible author:

Doz. Dr. JÜRGEN BENNDORF
Technische Universität Dresden
Sektion Wasserwesen
Bereich Hydrobiologie
Mommsenstr. 13
Dresden
DDR-8027

Limnologica (Berlin), 19 (1), 111—116, Juli 1988

Rezensionen

RUDOLF MANSFELDS **Verzeichnis landwirtschaftlicher und gärtnerischer Kulturpflanzen** (ohne Zierpflanzen). 2, neubearb. und wesentlich erweit. Aufl. Hrsgeg. von JÜRGEN SCHULTZE-MOTEL unter Mitarbeit der Autoren E. H. BENEDIX, R. FRITSCH, I. GREBENŠČIKOV, K. HAMMER, P. HANELT, J. KRUSE, H. I. MAASS, H OHLE, K. PISTRICK, A RIETH, J SCHULTZE-MOTEL und C TITTEL Zeichnungen: RUTH KILIAN und WOLFGANG KILIAN. Berlin: Akademie-Verlag 1986. Bd. 1: XVI + 577 S., 79 Abb.; Bd. 2. VI + 549 S, 74 Abb., Bd. 3: IV + 570 S., 108 Abb; Bd 4: IV + 302 S; Ln. M 320,—

1959 veröffentlichte RUDOLF MANSFELD das „Vorläufige(s) Verzeichnis landwirtschaftlich oder gärtnerisch kultivierter Pflanzenarten" als Beiheft 2 der „Kulturpflanze", in dem rund 1430 Arten niederer und hoherer Pflanzen erfaßt wurden, die garten- oder ackerbaulich als Nahrungs-, Futter-, Heil-, Öl-, Faser-, Gewürzpflanzen usw. kultiviert werden oder wurden. MANFELD schätze seinerzeit die Zahl der tatsächlich kultivierten Arten auf 1700—1800 und machte in seinem Vorwort auf die enormen Schwierigkeiten aufmerksam, die sich bei der Abfassung eines derartigen Kulturpflanzen-Handbuches ergeben, sei es aus Gründen der ungenügenden Kenntnis vieler Gruppen, sei es wegen der oft sehr unterschiedlich angewendeten Art-Konzeption oder sei es wegen des Fehlens von Zusammenstellungen der Kultur- und Nutzpflanzen für wichtige Gebiete, etwa in Zentralafrika, China oder Lateinamerika Trotzdem war der „MANSFELD" seinerzeit das einzig wirlich repräsentative Buch, das vorwiegend unter taxonomischen und nomenklatorischen Gesichtspunkten einen Überblick über die Kulturpflanzen bot. Es wurde durch das von JÜRGEN SCHULTZE-MOTEL verfaßte „Verzeichnis forstlich kultivierter Pflanzenarten". (Kulturpflanze, Beiheft 4, 1966) ergänzt. Inzwischen sind im russisch- und englischsprachigem Raum Publikationen ähnlicher Art erschienen, die jedoch eine andere Konzeption besitzen und z B. besonders die Verwendungszwecke oder die genetisch-züchterischen Aspekte betonen; taxonomische Belange treten in ihnen zuruck Das gilt auch für die von GUNTHER FRANKE und ARMIN FUCHS herausgegebenen „Nutzpflanzen der Tropen und Subtropen", die bei einer weit gesteckten Thematik inzwischen vier Auflagen erlebt haben, aber eben nur, wie aus dem Titel hervorgeht, einen Ausschnitt erfassen.

Das vorliegende Werk stellt die dringend notwendig gewordene Aktualisierung des MANSFELD-Verzeichnisses dar und wurde und wird als sogenanntes „MANSFELD-Projekt" nun seit rund 25 Jahren vom Zentralinstitut für Genetik und Kulturpflanzenforschung in Gatersleben betreut. Es verfolgt dasselbe Anliegen des alten „MANSFELD", wertet die Literatur der letzten 20 Jahre aus und berücksichtigt wesentlich stärker solche Nutzungsgruppen wie Futter- und Arzneipflanzen sowie solche, die als Gründüngungs- und Bodenbedeckungspflanzen, als Schattenbäume, Schutzhecken- oder Stützpflanzen in Hilfskulturen landwirtschaftlicher oder gärtnerischer Anpflanzungen Bedeutung besitzen. Darüber hinaus werden summarisch mehrere hundert weitere Arten behandelt, die bisher nur im Versuchsanbau oder in Kreuzungsprogrammen der Pflanzenzüchtung eingesetzt werden. Damit hat sich das Werk zu dem derzeit umfangreichsten Verzeichnis von Kulturpflanzen gemausert und umfaßt aus rund 230 Familien niederer und hoherer Pflanzen rund 4900 Sippen. Als wirtschaftlich wichtigste haben sich die Leguminosen mit 679 und die Gramineen mit rund 650 Arten herausgestellt. Ihnen schließen sich die Rosaceen und die Compositen (mit 238 bzw. 221 Arten) sowie die Euphorbiaceen, Labiaten und Solanaceen (150, 129 bzw 120 Arten) an Unter den Familien mit 50—100 Kulturarten sind vor allem die Liliaceen, Agavaceen und Palmen zu nennen Im allgemeinen wurden Gehölzsippen, die ausschließlich zu landeskulturellen Zwecken angebaut werden, ausgeschlossen. Die Sippen wurden nach ihrer Familienzugehörigkeit angeordnet, wobei die Reihenfolge der 12. Auflage des ENGLERschen „Syllabus der Pflanzenfamilien" folgt, die, das sei hier ausdrücklich betont, im Gegensatz zu vergleichbaren Werken, die Nutzpflanzen bereits gebührend berücksichtigte. In der Kennzeichnung der Gattungen und Arten sind die taxonomischen Basisdaten zur Benennung, Volksnamen, Angaben über das Wildareal bzw. das Anbaugebiet, die Nutzung der Sippe sowie in wichtigen Fällen Bemerkungen zu ihrer Phylogenie, Variabilität und geschichtlichen Ausbreitung enthalten. Die zunächst verwandten Wildarten, die als Ausgangssippen für Kulturarten zu gelten haben, sind aufgeführt Ein umfangreiches Literaturverzeichnis mit über 4600 Titeln ermöglicht ein weiterführendes Studium in verschiedene Richtungen. Die Literaturverweise sind am Ende der Behandlung der Gattung bzw. Art in codierter Form aufgeführt. Ein Teil der Arten (etwa 230) wird auch durch Abbildungen (insgesamt 241) belegt

Unsere bisherigen Ausführungen schließen sich eng an die Vorbemerkung des Buches an, weil diese konzis die wesentlichsten Eigenschaften in klarer und durchsichtiger Weise hervorheben. Es bleibt hier nur zu vermerken, daß die Bände 1 die Schizophyta, Phycophyta, Fungi, Pteridophyta, Gymnospermae sowie von den Angiospermae Casuarinaceae bis Leguminosa (1. Teil), 2 die Leguminosae (2. Teil), bis Verbenaceae und 3 die Labiatae bis Compositae sowie Alismetaceae bis Orchidaceae enthalten, während Band 4 eine Übersicht der Zahl der Kulturpflanzen-Arten in den einzelnen Familien, eine Liste der Neukombinationen, das Literaturverzeichnis sowie die Nachträge zur Literatur und zum Text, eine Übersicht der Arten nach dem Kulturzweck, Verzeichnisse der Volksnamen und der botanischen Namen sowie Berichtigungen enthält. Dem Herausgeber und den Autoren kann bescheinigt werden, daß sie mit dem neuen „MANSFELD" nicht nur eine empfindliche Lücke in der botanischen Literatur geschlossen haben, sondern daß sie eine einzigartige, hochaktuelle Dokumentation des von Landwirtschaft und Pflanzenzüchtung bei weitem noch nicht ausgeschöpften Potentials an bereits in Kultur genommenen Pflanzen geschaffen haben, die weltweit ihresgleichen sucht. Dieses internationale Standardwerk beweist, daß es durchaus möglich ist, theoretische Botanik in landwirtschaftliche oder gärtnerische Praxis umzusetzen: Dieses Buch ist nicht im oft geschmähten, sogenannten Elfenbeinturm der Wissenschaft angesiedelt, obwohl es aus dieser Quelle gespeist wird. Dem Verlag gilt höchstes Lob für die vorbildliche Ausstattung des Werkes, für dessen klaren Druck, der vor allem auch den unübertrefflichen Zeichnungen der Geschwister RUTH und WOLFGANG KILIAN zugutekommt (auch HANNA ROLOFF und ROLF PETRACK haben einige Zeichnungen beigesteuert). Hier hat die wissenschaftliche Zeichnung im deutschsprachigen Raum einen Gipfelpunkt erreicht — man sehe sich daraufhin z. B einmal die Abb. 6, 29, 35, 37, 42, 127, 172, 175, 176 oder 184 an. — Demgegenüber fallen leider die Wiedergaben der gewiß nicht schlechten Fotografien von G. SEIDEL, J. KRUSE, P. HILBERT, H KRASSMANN, N. FRANKE, GITTA TERPE und BRUNHILDE FOUQUET etwas ab Schade, daß der 4. Band, was das Papier angeht, nicht den Standard der ersten drei Bände fortführt. Das ändert aber nichts daran, daß hier eine große wissenschaftliche und verlegerische Leistung vorliegt, die international ein starkes positives Echo finden wird.

S. J. CASPER, Jena

GUSTAV HEGI. **Illustrierte Flora von Mitteleuropa.** Pteridophyta/Spermatophyta Hrsg.: H. J. CONERT, U. HAMANN, W. SCHULTZE-MOTEL, G. WAGENITZ. Band IV Angiospermae: Dicotyledones 2 Teil 1. Nachträge, Berichtigungen und Ergänzungen zum Nachdruck der 2. Auflage (1958—1963). Zusammengest. von W. SCHULTZE-MOTEL. Berlin, Hamburg: Verlag Paul Parey 1986. 70 S., 13 Abb. Brosch. DM 49,—

Dieses Heft bringt zunächst den Neudruck der Seiten 63, 196, 206, 213, 218, 221, 223, 398 und 494, bedingt durch die völlige Neubearbeitung der Verbreitungskarten von *Corydalis pumila*, *C. claviculata*, *Cardamine pratense*, *parviflora*, *trifolia*, *bulbifera*, *kitaibelii*, *heptaphylla*, *waldsteinii; Biscutella laevigata* incl. *B.* Ser. Laevigatae; *Crambe maritima*. Der Druck ist so eingerichtet, daß die neuen Seiten auf die alten aufgeklebt werden können Direkt als Anhang, daher mit Seiten 548 beginnend, ist der von W. SCHULTZE-MOTEL gestaltete Abschnitt „Nachträge, Berichtigungen und Ergänzungen zum Nachdruck der 2. Auflage von Band IV/1 (1958—1963)" gedacht Hier finden sich zahlreiche wichtige Hinweise Besonders wesentlich sind die Ergänzungen bei *Papaver*, *Corydalis*, *Erysimum*, *Rorippa*, *Cardamine*, *Arabis* und *Draba*. Neue Arealkarten: *Cardamine pentaphyllos*, *C. glanduligera*, *C. enneaphyllos*. Sie sind, wie die bereits oben genannten, vor allem H. D. KNAPP zu verdanken. Nach wie vor verwunderlich bleibt die Reserve, die der Bearbeiter gegenüber den *Thlaspi*-Bearbeitungen von F. K. MAYER an den Tag legt. Druck und Ausstattung sind hervorragend. Eine wertvolle Bereicherung des „HEGI".

S. J. CASPER, Jena

HELMUT MÜHLBERG: **Vermehrung der Aquarienpflanzen.** Auswahl — Fortpflanzung — Zucht Leipzig, Jena, Berlin: Urania-Verlag 1986 79 S., 32 Abb., 2. Tab. Kst. M 9,80

Der Hallenser Wasserpflanzenspezialist, dem wir bereits „Das große Buch der Wasserpflanzen" (1980) verdanken, vermittelt in diesem handlichen Büchlein die erfolgreiche Praxis des Wasserpflanzengärtners und gibt Anleitungen und Tips zur generativen und vegetativen Vermehrung der Aquarienpflanzen. Es ist ja heute so, daß im Aquarium nicht nur Fische sich in großer Zahl drängen, sondern zu einem schönen Aquarium gehört eine ansprechende Bepflanzung. Viele der beliebtesten Aquarienpflanzen stammen aus tropischen und subtropischen Breiten. Für die meisten Aquarianer ist daher die Vermehrung von Aquarienpflanzen von besonderer Bedeutung. Leider lassen sich nicht alle problemlos vermehren. Hier tritt MÜHLBERGS „Vivaristik-Ratgeber in Farbe" — so bezeichnet der Verlag eine Reihe von Büchern, die Einzelgebiete der Vivaristik behandeln sollen — in Aktion. Er will den Bemühungen des Aquarienfreundes durch seine praktischen Ratschläge und durch die Vermittlung theoretischer Grundlagen zum Erfolg verhelfen.

Das Büchlein ist in vier Hauptabschnitte gegliedert. Der erste behandelt die vegetative Vermehrung an zahlreichen einschlagigen Beispielen (Vermehrung durch Ableger, durch Stecklinge, durch Kriechtriebteilung, durch Rhizomteilung und mittels flutender Sproßstücke). Eine beachtliche Rolle spielt die Vermehrung durch Adventivpflanzen, wofür aus den Gattungen *Ceratopteris*, *Echinodorus*, *Aponogeton* und *Hygrophila* charakteristische Beispiele vorgeführt werden Kurz wird auch der Gewebekultur am Beispiel von *Cryptocoryne* gedacht. Der zweite Abschnitt beschäftigt sich mit der generativen Vermehrung. Hier ist es zu begrüßen, daß eine Reihe grundsätzlicher Kenntnisse über den Bau von Blüte, Frucht und Samen sowie über Bestäubung, Befruchtung und Samenbildung vermittelt werden. Im dritten Abschnitt wird die generative Vermehrung der Froschbißgewächse, der Gattungen *Aponogeton*, *Cryptocoryne* und *Anubias* sowie der Farnpflanzen (hier der Wasserfarne und der Brachsenkräuter) praxisnah erörtert. Der vierte Abschnitt über die Züchtung bei Aquarienpflanzen besteht eigentlich nur aus ein paar Notizen über Bedeutung und Möglichkeiten, gewinnt aber durch den Anhang von 2 Tabellen, die in übersichtlicher Form übliche sowie zusätzliche oder biologisch interessante Vermehrungsformen bei den wichtigsten Gattungen auflisten.

Die Darstellung ist vorbildlich, die Diktion einfach, sachbezogen und sehr gut lesbar. Hervorragende Zeichnungen überzeugen durch ihre Aussagekraft; die 10 Farbfotos sind von hoher Qualität. Da auch Papier und Druck höchsten Ansprüchen genügen, kann das Buch unter ähnlichen Publikationen eine Sonderstellung beanspruchen Populärwissenschaft, wie sie sein soll.

S. J. CASPER, Jena

RAINER KOCH **Datenspeicher Umweltchemikalien.** Eigenschaften, Toxizität, Umweltverhalten. Berlin VEB Verlag Volk und Gesundheit 1986 332 S., 18 Tab., 10 Abb. Brosch. M 41,—

„Chemie gibt Brot, Wohlstand und Schönheit", so wurde vor etwa 30 Jahren emphatisch formuliert.

Es ist unzweifelhaft, die Chemie hat besonders in den modernen Industriestaaten einen Stellenwert erreicht, der uns im einzelnen kaum noch bewußt wird Was wären z. B. die Landwirtschaft ohne chemische Düngemittel und Pestizide? Auch Synthetiks und Kosmetika sind aus dem täglichen Leben nicht mehr wezudenken.

Andererseits haben wir inzwischen zu Kenntnis nehmen müssen, daß das Klischee von Brot, Wohlstand und Schönheit so einfach nicht ist Jeder, der Augen, Ohren und Nase offen hält, weiß von der Bedrohnung, die von der Chemie bzw. von der Anwendung chemischer Kenntnisse in der industriellen Praxis ausgehen kann. 1983 kannte man immerhin 6 Millionen chemische Verbindungen, von denen 60000—70000 kommerziell wichtig waren Es ist unvermeidbar und zum Teil auch gewünscht, daß ein mehr oder weniger großer Teil dieser Stoffe direkt auf den Menschen einwirkt oder indirekt über vielfältige Wege durch die Umwelt an ihn gelangt. Deswegen ist ihre Bewertung im Hinblick auf die menschliche Existenz und die seiner Umwelt von größtem Interesse.

RAINER KOCH hat etwa 120 dieser Verbindungen in einen Katalog aufgenommen und bewertet. Die Auswahlkriterien waren u. a. Produktionswege, Anwendungsmenge, Wirkungsparameter, Umweltverhalten und Toxizität. Die Charakterisierung der Substanzen erfolgt anhand von sieben Hauptkriterien, die in Anlehnung an internationale Standards gewählt wurden.

Neben der qualitativen Problematik gibt es eine Reihe offener Fragen, die Merkmale einzelner Stoffe und ihre Bewertung betreffend.

So ist z. B. die Toxizität in den meisten Fallen an Tieren, Mikroorganismen oder Zellkulturen ermittelt worden. Die Übertragung der Versuchsergebnisse auf den Menschen erfordert deshalb gesonderte Überlegungen. Der Autor schreibt dazu: „Jede Interpretation toxikologischer Daten muß .. beachten, daß toxikologische Experimente nahezu immer mit nicht umweltadäquaten Stoffmengen und unter Modellbedingungen durchgeführt werden." Die Existenz eines Schwellenwertes wird angenommen. Allerdings zeigt der Mensch oft größere Empfindlichkeiten als die Versuchstiere Die dadurch notwendigen Sicherheitsfaktoren sind meistens empirisch festgelegt, weil notwendige Kenntnisse fehlen. In manchen Fällen ist es sogar zweifelhaft, ob Grenzwerte überhaupt berechtigt sind. Ein wichtiges Problem des Umweltverhaltens chemischer Verbindungen ist die Stofftransformation, die entgiftend aber auch giftend wirken kann.

In einem Glossar definiert der Autor eine Reihe von Begriffen, die die Lektüre und das Verständnis seines Buches erleichtern sollen Neben anderen, einleuchtenden Erklärungen erscheint auch das Begriffspaar „Ökotoxikologie, ökotoxikologisch". Es bleibt Geheimnis des Autors, warum zu seiner Beschreibung das Bio- nicht aber das Ökosystem benutzt wird Ebenfalls hier, an recht verborgener Stelle, findet sich die Definition der Begriffe „Umweltchemikalie" bzw „Umweltschadstoff". Hier müßte nach meiner Auffassung schärfer formuliert und besser getrennt werden.

Den Hauptteil des Buches bildet die alphabetisch geordnete Folge der Substanzen und ihre Bewertung nach den oben genannten Gesichtspunkten Neben weniger bekannten Verbindungen finden sich auch solche Stoffe, die bereits eine traurige Berühmtheit erlangt haben, wie etwa Cyanide, DDT, polychlorierte Biphenyle, polycyclische aromatische Kohlenwasserstoffe und Methylquecksilber.

Insgesamt ist mit dem „Datenspeicher Umweltchemikalien" ein wichtiges Buch erschienen, dem nicht nur schlechthin weite Verbreitung, sondern Kenntnisnahme und praktische Reaktion zu wünschen sind.

P. KASPRZAK, Neuglobsow

KLAUS-DIETER RADEMACHER & KLAUS-DIETER KOSS · **Wassergefährdende Stoffe.** Vorschriften und Erläuterungen. Berlin, Heidelberg, New York, Tokyo: Springer-Verlag 1986. 366 S., 13 Abb., 6 Tab. Brosch. DM 240,—

In einem modernen Industriestaat mit angespannter Wassersituation müssen zum Schutze des Wassers Gesetze und Bestimmungen erlassen werden. Für einen breiten Anwenderkreis, besonders für den Mann der Praxis, wurde in leicht handhabbarer Form eine Sammlung einschlägiger Gesetze und Verordnungen der BRD zusammengestellt. Wie groß die Zahl ist, zeigt das umfangreiche Inhaltsverzeichnis am Ende der Broschüre (hier könnte die Angabe der entsprechenden Seitenzahl — anstelle der Kapitelnummer — die Handhabung des Buches erleichtern).

Das Buch ist in 8 Kapitel eingeteilt, die folgende Schwerpunkte erfassen: Wasserrechtliche Bestimmungen, Verordnungen über Anlagen für wassergefährdende Stoffe und eine alphabetische Auflistung der häufigsten wassergefährdenden Elemente und Verbindungen mit einer Einordnung in die Gefährdungsklasse. Es wird auch auf die Zulassung von Fachbetrieben, auf gewerbliche Bestimmungen, Alarmeinrichtungen und Strafbestimmungen eingegangen.

Es ist zu wünschen, daß eine weite Verbreitung der Kenntnisse der Gesetze und Verordnungen zum Schutz der Umwelt und zur Erhaltung und Sanierung der Gewässer beitragen möge.

G. PROFT, Jena

MOHIUDDIN MUNAWAR & JOHN FRANCIS TALLING (eds.): **Seasonality of freshwater phytoplankton.** A global perspective (Developm. in Hydrobiol. 33). Dordrecht, Boston, Lancaster: Dr. W. Junk Publishers 1986. VIII + 236 S., 136 Abb., 40 Tab. Dfl. 200.— (US $ 95.50, £ 64.25)

Dieser „Sonderdruck" des Bandes 138 (1986) der Hydrobiologia versucht, einen weltweiten Überblick über die jahreszeitliche Entwicklung des Phytoplanktons in den Binnengewässern zu geben. Die Herausgeber fanden anläßlich des SIL-Kongresses in Lyon (August 1983) genügend Mitstreiter für ihr Unternehmen, das sie angesichts des hohen Kenntnisstandes vor allem in den gemäßigten Breiten für längst überfällig hielten Insgesamt trugen 25 anerkannte Phytoplanktonforscher aus der BRD, aus Österreich, Großbritannien, der UdSSR, den USA, aus Kanada, Israel, Indien, Peru und Neuseeland zum Gelingen des Bandes bei, der bei vielen Limnologen auf reges Interesse stoßen wird

Wissenschaftshistorisch gesehen ist die raum-zeitliche Phytoplanktonanalyse in der nördlich gemäßigten Zone zu Hause. Folgerichtig sind Mittel- und Westeuropa durch vier Beiträge vertreten U SOMMER vergleicht die Periodizität des Phytoplanktons des Bodensees mit der anderer tiefer Seen in Mitteleuropa. M DOKULIL und C SKOLAUT diskutieren die Phytoplanktonsukzession am Beispiel des Mondsee (Österreich), eines tiefen, geschichteten Sees. A. E. BAILEY-WATTS richtet das Augenmerk auf die jahreszeitlichen Veränderungen der Größe der Phytoplankter im schottischen Loch Leven. C. S. REYNOLDS manipuliert die Phytoplanktonperiodizität in großen Kompartimenten im Blelham Tarn (English Lake District). Spezielle Aspekte unter Berücksichtigung der besonderen Bedingungen in den nördlichen Breiten beleuchten N. A. PETROVA (*Melosira*-Plankton im Ladoga- und Onega-See) und R. G. SHEATH (Phytoplankton in nördlichen Tundra-Weihern). Es ist schade, daß für diese Fragestellung kein Autor aus Skandinavien gewonnen werden konnte. Die großen Seen des nordamerikanischen, klimatisch kontinental geprägten Kontinents werden unter Einschluß des Pico- und Ultraplanktons (M. MUNAWAR & I. F. MUNAWAR: North American Great Lakes) und des Sediment-Wasseraustausches (R. E. HECKY, H. J. KLING, G. J. BRUNSKILL) übersichtlich und informativ vorgestellt. Als subtropischer See wird der Lake Kinneret (Israel) auf das Verhalten des Phytoplanktons hin untersucht (U POLLINGHER). Einen Überblick über die verschiedenen jahreszeitlichen Muster auf dem afrikanischen Kontinent gibt J F TALLING, und J. KALFF und S. WATSON vergleichen die Phytoplanktondynamik in zwei tropischen afrikanischen Seen mit der in der gemäßigten Zone. Erfahrungen unter den Bedingungen des Monsunklimas (Sudindien) gibt A. R ZAFAR weiter. Auf Langzeituntersuchungen kann sich W. M. LEWIS, jr., stützen, der die Phytoplanktonsukzession im Tieflandsee Valencia in Venezuela intensiv erforscht hat. Die Zusammenhänge von physikalischen und chemischen Prozessen mit der spezifischen raum-zeitlichen Ausprägung der Phytoplanktonzönose behandeln P. J. RICHERSON, P. NEALE, W. WURTSBAUGH, R. ALFARO und W. VINCENT Den Abschluß bildet die Darstellung der Phytoplanktonentwicklung der Waitaki lakes in Neuseeland durch H. C DUTHIE und V. M. STOUT, die zeigen, daß in den gemäßigten Breiten der Sudhemisphäre nur wenige brauchbare Fakten ermittelt worden sind. Ost- und Südostasien sind nicht repräsentiert.

Es muß betont werden, daß alle Artikel auf einem hohen wissenschaftlichen Niveau stehen und praktisch alle theoretischen Vorstellungen, Methoden und Techniken enthalten, die gegenwärtig für die Analyse der jahreszeitlichen Phytoplanktondynamik relevant sind. Insofern kann der Sammelband richtungsweisend wirken Die Textdarstellung ist knapp und durchsichtig. Sie wird von zahlreichen Abbildungen und Tabellen gut unterstützt. Druck und Ausstattung sind fast durchweg gut.

S. J. CASPER, Jena

WINFRIED LAMPERT (ed.): **Food limitation and the structure of zooplankton communities.** Proceedings of an International Symposium held at Plön, W. Germany, July 3—13, 1984. Stuttgart: E. Schweizerbart'sche Verlagsbuchhandlung (Nägele und Obermiller) 1985. 497 S., 207 Fig., 82 Tab. DM 188,—

Als sich die Limnologie zu Beginn unseres Jahrhunderts zu einer eigenständigen biologischen Disziplin entwickelte, galt die Aufmerksamkeit von Anfang an auch dem Zooplankton. Das zeigen Arbeiten von BIRGE, JUDAY und NAUMANN. Doch dann ließ das Interesse am Zooplankton nach, und sowohl die spätere Eutrophierungsforschung als auch das Internationale Biologische Programm änderten diese Grundrichtung nur teilweise, denn im Vordergrund stand die Wechselwirkung zwischen den Nährstoffen und den Primärproduzenten, vor allem den Planktonalgen.

Es ist deswegen nicht verwunderlich, daß einige Arbeiten des tschechischen Limnologen HRBÁČEK und eine Veröffentlichung der beiden Amerikaner BROOKS & DODSON zu Beginn der sechziger Jahre, die sich mit dem Einfluß der Nahrungskettenstruktur auf das trophische Geschehen in stehenden Gewässern befaßten, entweder nicht bemerkt wurden oder schon bald wieder in Vergessenheit gerieten.

Diese Situation hat sich seit 1975 drastisch geändert, als eine Gruppe amerikanischer Limnologen um SHAPIRO Möglichkeiten aufzeigte, die Nahrungskettenstruktur für die Zwecke der Gewässergütesteuerung zu nutzen.

Seit dieser Seit erlebt die Zooplanktonforschung eine wahre „Renaissance", und letztendlich ist der vorliegende Symposiumsband ein, wenn auch sehr indirektes, Produkt dieser Geschichte.

54 Teilnehmer aus 14 Ländern hielten zu 8 Themenkomplexen 44 Vorträge. Der Herausgeber, W. LAMPERT, legte das Schwergewicht des Symposiums auf den strukturbildenden Einfluß des Futterangebotes auf die Zooplanktongemeinschaft. Damit soll die eigenständige Bedeutung von Futterquantität und -qualität im Vergleich zur Rolle der Zooplanktonräuber betont werden. Hierdurch wurde es auch möglich, ausgewählte Fragen der Ernährungsbiologie mariner Copepoden zu integrieren, deren Populationsstruktur offensichtlich noch mehr als die limnischer Zooplankter vom Futterangebot gesteuert wird.

Die Themenkomplexe befassen sich mit der Frage nach der Futterlimitation des Zooplanktons, den Mechanismen der Futterbeschaffung bei geringem Angebot, der Reaktion des Zooplanktons auf wechselnde Futterkonzentrationen, den Überlebens- und Wachstumsstrategien bei geringem Futterangebot, der Populationsdynamik als Indikator der Futterlimitation, den Auswirkungen auf den Lebenszyklus, dem Verhältnis von Futterlimitation und Räubereinfluß und schließlich der Konkurrenz bei begrenztem Futterangebot.

Im Anschluß sollen einige Arbeiten etwas naher vorgestellt werden, ohne damit eine allgemeine Wertung vornehmen zu wollen

GULATI, SIEWERTSEN & POSTEMA fanden in geschichteten, mesotrophen Seen der Niederlande einen deutlichen Einfluß des Zooplanktons auf die Sestonkonzentration. Dabei kommt es neben der quantitativen zusätzlich zu einer qualitativen Limitation, die in der Anreicherung des Sestongemisches mit schlechtverwertbarem Detritus durch die Zooplankter selbst besteht. In hocheutrophen Flachseen bestand keine Beziehung zwischen Zooplankton- und Sestonmasse.

Hervorzuheben ist die Arbeit von BRENDELBERGER, der jüngste Diskussionen über den Filtrationsmechanismus von Cladoceren aufgreift und experimentell testet. Danach wirken die Filterbeine der Tiere als einfache Siebe und die Größe der zurückgehaltenen Partikel läßt sich anhand der „Maschenweite" der Filterborsten vorhersagen.

MAYZAUD & MAYZAUD untersuchten die Aktivität von Verdauungsenzymen maritimer, calanoider Copepoden. Es zeigte sich, daß sie sowohl von der Futterqualität und -quantität, als auch von den jeweiligen Energiebedürfnissen der Tiere geregelt wird.

LAMPERT & MUCK befaßten sich mit dem häufigen empirischen Befund einer Ablösung von *Diaptomus* durch *Daphnia* im Verlaufe der Eutrophierung Sie konnten mit Hilfe von Laborexperimenten eine bessere Anpassung von *Eudiaptomus gracilis* an niedrige und stark fluktuierende Futterkonzentrationen im Vergleich zu *Daphnia hyalina* und *Daphnia galeata* nachweisen. Die Grundlage der größeren Widerstandfähigkeit von *Eudiaptomus gracilis* ist ein „sparsamerer" Umgang mit der Energie

Schließlich sei noch die Arbeit von BENNDORF & HORN genannt, die sich mit dem Verhältnis von Futterlimitation und Räuberdruck auf die Struktur der Zooplanktongemeinschaft beschäftigen. Sie fanden, daß die Zusammensetzung und Menge des Futters strukturbildend in Bezug auf die vorherrschenden Ernährungstypen (Mikro-, Makrofiltrierer, Beutegreifer) wirkt. Diese Kontrolle ist unabhängig von der Körpergröße. Die Art und Intensität des Räubereinflusses regulieren dagegen die Größenzusammensetzung der Zooplanktongemeinschaft, modifiziert durch die Fähigkeit der Beutetiere, dem Räuberzugriff zu entkommen (z. B. *Daphnia*, *Eudiaptomus*).

P. KASPRZAK, Neuglobsow

H. C. TORNO, J. MARSALEK & M. DESBORDES (eds.): **Urban Runoff Pollution.** Proceedings of NATO Advanced Research Workshop on Urban Runoff Pollution, held in Montpellier, France, Aug. 1985. Berlin, Heidelberg, New York: Springer-Verlag 1986. 893 S., Geb. DM 290,–

Bei den Bemühungen um Eindämmung der Gewässerbelastung aus sogenannten „punktförmigen Quellen" zeigt es sich vor allem in hochindustrialisierten Ländern immer wieder, daß nur begrenzte Sanierungserfolge möglich sind, wenn die flächenhaften Stoffeinträge unberücksichtigt bleiben. Neben den „diffusen" Stoffeinträgen aus Gebieten mit intensiver Landwirtschaft trifft dies insbesondere für die Einträge aus Stadtentwässerungssystemen zu, insbesondere in urbanen Ballungszentren. Das Zusammentreffen von 46 Teilnehmern vor allem aus den USA (10), aus Kanada (8) und aus Frankreich (8) in Montpellier sollte dazu beitragen, den gegenwärtigen Kenntnisstand auf diesem Gebiet breitenwirksamer zu gestalten und Schlußfolgerungen für weitere notwendige Forschungs- und Entwicklungsarbeiten ableiten zu können.

Die vorliegenden Proceedings enthalten 29 Beiträge, die zehn Themengruppen zugeordnet wurden: Verschmutzungsaspekte von Stadtflächenabflüssen (3) — Beispielsuntersuchung zur Qualität von Stadtflächenabflüssen (4) — Deterministische Modellierung der Abflußprozesse (2) — Statistisch gestützte Modellierung der Abflußprozesse (3) — Herkömmliche Verschmutzungseinflüsse auf die Vorfluter (2) — Toxische Belastungseinflüsse der Vorfluter (2) — Modellierung der Abflüsse in den Vorflutern (4) — Bewirtschaftung/Behandlung der Qualität des abfließenden Wasser (5) — Echtzeitkontrolle (3) — Notwendige weitere Forschungsarbeiten (1).

Die meisten der Beiträge sind mit ausführlichen Literaturnachweisen versehen, was dem auf diesem Gebiet Suchenden entgegenkommt und einem wichtigen Anliegen des Symposiums, der Verbreitung des gegenwärtigen Kenntnisstandes, gerecht wird.

Einem anderen Anliegen des Arbeitstreffens — der Ableitung von Schlußfolgerungen für weitere Arbeiten — dienen sowohl die Detaildiskussionen zu den jeweiligen Themengruppen, als auch eine abschließende Gesamtdiskussion auf der Basis einer — auf der Meinungsgrundlage der Teilnehmer erstellten — Prioritätenliste. Am häufigsten wurden dabei Untersuchungen zum wachsenden Verständnis der Prozesse und Mechanismen — einschließlich der entsprechenden mathematischen Modellierung — der Schmutz- bzw. Schadstoffablagerung und des Transportes gefordert. Im Vergleich zu den Aussagen des Internationalen Symposiums zur Stadtflächenhydrologie im Juni 1983 in Baltimore (USA) zeigen sich in den Prioritäten vor allem veränderte Standpunkte zur stärkeren Erfassung der Auswirkungen auf die Vorfluter. Nach wie vor muß jedoch der Mangel an zuverlässigen Daten über die äußerst dynamischen, vielfältig verknüpften Prozeßverläufe und Komplexwirkungen als Haupthindernis für effektivere Steuerungs- und Bewirtschaftungsverfahren genannt werden.

U. GRÜNEWALD, Dresden

BRIAN E. S. GUNNING & MARTIN W. STEER: **Bildatlas zur Biologie der Pflanzenzelle.** Aus dem Englischen übersetzt von KURT DIETER ZAAR. 3., durchgesehene Aufl. Stuttgart, New York: Gustav Fischer Verlag 1986. VI + 103 S., 49 Taf. Kt. DM 34,–

Die deutsche Ausgabe des 1975 bei EDUARD ARNOLD (Publishers) Ltd in London erstmals erschienenen „Sonderdrucks" des Tafelteils von GUNNINGS & STEERS „Ultrastructure and the Biology of Plant Cells" (vgl. Rezensionen in Limnologica **11.** S. 182 bis 1983; 1977) ist bisher unter dem Titel „Biologie der Pflanzenzelle. Ein Bildatlas" (1. Auflage 1977) herausgebracht worden. Über 200 hervorragende licht- und elektronenmikroskopische Abbildungen pflanzlicher Gewebe werden eindrucksvoll und aussagekräftig dargestellt. Den Abbildungen entspricht ein zwar knapper, aber alles Wichtige beinhaltender Text, der auch auf die jeweiligen Präparationsmethoden eingeht. Ausstattung und Druck der praktisch unveränderten Neuauflage genügen allen Ansprüchen Lehrer und Studenten der Biologie werden das Buch mit Gewinn benutzen können.

S. J. CASPER, Jena

KURT KRAMMER: **Kieselalgen, Biologie, Baupläne der Zellwand, Untersuchungsmethoden** (Kosmos-Handbücher für die praktische naturwissenschaftliche Arbeit). Stuttgart: Franckh'sche Verlagshandlung 1986. 140 S., 243 Abb Kart. DM 39,80

Die Kieselalgen haben seit jeher durch die Schönheit ihrer Formen und die relativ leichte Präparierbarkeit unter den Naturfreunden Liebhaber gefunden. Natürlich bedurften diese zur Untersuchung des Mikroskopes, und bis in die fünfziger Jahre unseres Jahrhunderts hinein reichte diese Technik auch vollständig hin, um die reizvollen mikroskopischen Objekte näher kennenzulernen. Noch 1956 veröffentliche der Altmeister der Kieselalgenkunde in Mitteleuropa, FRIEDRICH HUSTEDT, übrigens ebenfalls im Kosmos-Verlag, ein Bändchen „Kieselalgen (Diatomeen)", das in souveräner Weise alles zusammenfaßte, was bis dahin mit Hilfe des Lichtmikroskopes an der künstlich „fossilierten" Kieselalgenschale erkannt werden konnte. Es war gleichsam ein Abgesang auf das Zeitalter des Lichtmikroskopes, weil nunmehr die intensive Verwendung von Transmissions- und Raster-Elektronen-Mikroskopen die Erkenntnisse über den Bau der Diatomeenschale, über ihre Morphogenese und geschlechtliche Fortpflanzung in ungeahnter Weise voranbrachte. Zwar sind die Diskussionen über die Deutung von morphologischen Strukturen in licht- und elektronenmikroskopischen Bildern noch immer vielfach kontrovers, doch setzt sich stetig die Erkenntnis durch, daß ohne das Elektronenmikroskop „nichts mehr zu machen ist".

Was in den letzten 25 Jahren geleistet worden ist, stellt der Autor, der sich selbst seit über 30 Jahren mit Kieselalgen beschäftigt, in überzeugender Weise vor Was seine Darstellung besonders auszeichnet, ist der direkte Vergleich von raster-elektronen-mikroskopischen Aufnahmen mit lichtmikroskopischen Bildern auf 45 exzellenten Fototafeln.

Im einzelnen behandelt die Darstellung konzis und dennoch ausführlich genug den allgemeinen Aufbau der Zellwand, die Methoden zur Untersuchung von Kieselalgen sowie die Baupläne und Strukturen der Diatomeenzellwand am Beispiel ausgewählter Gattungen. Nur knapp werden Fragen wie vegetative Vermehrung, geschlechtliche Fortpflanzung und Auxosporenbildung, plasmatische Strukturen und Morphogenese der Zellwand sowie Überlebensstrategien und Ökologie erörtert. Der Leser findet im Buch nur zwei Zeichnungen. Der Autor hat sich damit klar für die Fotografien entschieden, was er eingehend begründet. Auch wenn man seinen Argumenten nicht ganz folgen können wird, überzeugen die Fotos durch ihre Qualität. Nützlich ist ein Glossar, das eine Terminologie der verkieselten Schalenstrukturen bietet. Literaturverzeichnis und Register bilden den technischen Apparat des hervorragend gedruckten Buches, das ohne Zweifel unentbehrlich für jeden ist, der sich mit Kieselalgen befaßt, gleichgültig ob er sich Fachmann oder Liebhaber nennt. Im Buch finden sich viele Hinweise auf das vom Autor zusammen mit LANGE-BERTALOT (1986) verfaßte Bestimmungswerk „Bacillariophyceae" (Band 2/1 der Süßwasserflora von Mitteleuropa), zu dem dieses Kosmos-Handbuch eine wertvolle Ergänzung darstellt.

S. J. CASPER, Jena

ROGER ECKERT: **Tierphysiologie.** Mit Beiträgen von DAVID RANDALL. Übersetzt und bearbeitet von RAIMUND APFELBACH unter Mitarbeit von E. WEILER, R. KÄSTNER und I. FEUERBACHER. Stuttgart: Georg Thieme Verlag 1986. 697 S., 728 Abb., 47 Tab. Kart. DM 69,—

Das vorliegende Buch der Tierphysiologie ist die übersetzte und bearbeitete 2 Auflage der bei W. H. Freeman & Comp. (San Francisco) 1983 erschienen Originalausgabe „Animal Physiology — Mechanisms and Adaptions".

Es ist in 5 Abschnitte untergliedert und umfaßt insgesamt 16 Kapitel. Die ersten beiden Kapitel behandeln physikalische und chemische Grundlagen physiologischer Abläufe. Im 3 und 4. Kapitel werden Fragen der Zellenergetik und der Regulation des intrazellulären Milieus (Enzymatik und Energetik, Permeabilität und Transport) erörtert. Die Kapitel 5 bis 8 behandeln erregbare Membranen, Nervensignale, sensorische Mechanismen und die Funktionsweise des Nervensystems. In den Kapiteln 9 und 10 werden Fragen der Kontraktilität und Motilität diskutiert. Die Kapitel 11 bis 16 sind den Systemen gewidmet, die für die Homöostase und Versorgung des inneren Milieus verantwortlich sind (chemische Botenstoffe, Osmoregulation und Exkretion, Blutzirkulation, Gasaustausch, Ernährung und Verdauung, Energiehaushalt).

Das Lehrbuch enthält meist zweifarbige Abbildungen, und jedes der 16 Kapitel beschließt mit einer Zusammenfassung.

E. HENTSCHEL, Jena

Veröffentlichungen des Geobotanischen Institutes der ETH, Stiftung Rübel, Zürich (Heft 89). TH. B. EGLOFF: Auswirkungen und Beseitigung von Düngungseinflüssen auf Streuwiesen. Zürich 1986. 183 S., 79 Abb., 36 Tab. Brosch. SFr. 42,—

Im vorliegenden Band — der Dissertation des Autors — wird ein für die praktische Naturschutzarbeit grundlegendes Thema behandelt. Die Arbeit geht zwei wesentlichen Fragen nach, den Auswirkungen der Eutrophierung auf Streuwiesen und den Möglichkeiten einer Oligotrophierung bzw Regeneration eutrophierter Streuwiesen durch gezielte Nutzungs- bzw Pflegeeingriffe.

Zum erstgenannten Problemkreis wurden Düngungsversuche durchgeführt und besonders hinsichtlich der Eutrophierungsindikatoren und der für das Pflanzenwachstum primär limitierend wirkenden Nährstoffe ausgewertet. Die Ergebnisse der Versuche verdeutlichen auch die Vielfalt möglicher Entwicklungen der Vegetation nach Düngergabe.

Im Rahmen der Regenerationsexperimente wurden die Auswirkungen verschiedener Schnittregime analysiert. Neben den Fragen zur günstigsten Schnitthäufigkeit und zum Zeitpunkt sowie zur Höhe des Nährstoffentzuges durch Schnitt sind auch die Wirkungen auf einzelne Arten genauer dargestellt worden.

Abschließend werden „Richtlinien zur Erhaltung artenreicher Riedwiesen" unterbreitet. Sie lassen sich folgendermaßen zusammenfassen: Keine Änderungen des Wasserhaushaltes; keine Düngung; einschürige Mahd im Spätherbst (oder Winter) mit Beräumung des Mahdgutes; Ausweisung von Pufferzonen, floristische Überwachung, gegebenenfalls spezielle Eingriffe. Der Schutz noch vorhandener wertvoller Bestände ist bedeutend günstiger einzuschätzen als die Aushagerung bereits eutrophierter Wiesen.

Die zahlreichen Ergebnisse wurden mathematisch gut aufbereitet und auf der Grundlage einer umfangreichen Literaturauswertung (214 Literaturstellen) ausführlich diskutiert. Durch eine straffere Gliederung und bessere Stilistik wäre es jedoch möglich gewesen, die Aufmerksamkeit des Lesers stärker auf das Wesentliche zu konzentrieren. Einige weitere kleinere Mängel schmälern jedoch keineswegs den hohen Gesamtwert der Arbeit für die Naturschutzpraxis. Nicht zu Unrecht hat der Autor seine Arbeit „den Lebewesen der Rieder und Moore" gewidmet, die nicht nur sein Untersuchungsobjekt darstellten, sondern ihn auch emotional stark angesprochen haben, so daß seine Engagement für deren Erhaltung deutlich spürbar wird. Es wäre zu wünschen, daß die als Dissertation begonnenen Untersuchungen langfristig fortgesetzt werden.

W. WESTHUS, Jena

Veröffentlichungen des Geobotanischen Institutes der ETH, Stiftung Rübel, Zürich (Heft 90). OTTO WILDI: Analyse vegetationskundlicher Daten. Zürich 1986. 266 S., 29 Tab., 46 Abb. Brosch. SFr. 36,—

Heft 90 der bekannten Züricher Schriftenreihe enthält eine eingehende Darstellung von Theorie und Einsatz statistischer Methoden in der Vegetations- und Standortkunde. In 9 Kapiteln und einer abschließenden Diskussion werden die einschlägigen Methoden und ihre theoretischen Grundlagen vorgestellt und an Beispielen erläutert. Für die erfolgreiche Anwendung statistischer Verfahren in Forschung und Praxis hat der Verfasser ein eigenes Paket von Computerprogrammen entwickelt, welches einige der hier beschriebenen Verfahren zu realisieren erlaubt. Er verweist im übrigen auf weitere wichtige bereits entwickelte und publizierte Programme Dennoch muß er am Ende bekennen, daß es selbst bei Verwendung der besten Computerprogramme möglich ist, klare Strukturen zu übersehen und richtige Ergebnisse als falsch zu betrachten. Denn die Vegetationskunde gehört zu den sogenannten „weichen" Wissenschaften, sie beschäftigt sich mit Gegebenheiten, die sehr komplex und in der Regel schwer erkenn- und nachvollziehbar sind

Das vorliegende Heft ist wichtig für alle diejenigen Vegetationsforscher, welche bei ihrer Arbeit Computer einsetzen wollen und denen die entsprechenden Geräte zur Verfügung stehen.

H.-D. KRAUSCH, Potsdam

ULRICH LEHMANN: **Paläontologisches Wörterbuch.** 3., überarb. u erw. Aufl. Jena: VEB Gustav Fischer Verlag 1986. 440 S., 112 Abb 4 Tafeln DDR M 23,50/Ausland DM 29,—

Die Paläontologie ist eine jener jungen Naturwissenschaften, die sich zu Beginn des vorigen Jahrhunderts verselbständigten und im Laufe ihrer kurzen Geschichte eine ungeheuere Materialfülle anhäuften, was sowohl die Fakten als auch die Theorien anbelangt. Für die Evolutionsforschung in der Biologie, die praxiswirksame Stratigraphie in der Geologie, die neueren Teildisziplinen der Paläogeographie und Paläoökologie sowie für eine auf naturwissenschaftlichen Grundlagen fußende wissenschaftliche Weltanschauung ist diese faszinierende Wissenschaft unentbehrlich geworden. Sie stellt die Verbindung zwischen den Geowissenschaften auf der einen und den Biowissenschaften auf der anderen Seite her. Ihre Fachsprache ist, wie der Autor betont, sehr weitgespannt und reich an Begriffen und Namen aus beiden Wissenschaftszweigen, ein Umstand, der selbst Fachleuten bisweilen den Zugang zur wissenschaftlichen Paläontologie zu verwehren droht.

Eine dritte überarbeitete und erweiterte Auflage eines paläontologischen Taschenwörterbuches, der ersten Ausgabe eines solchen Buches in der Deutschen Demokratischen Republik, ist eine begrüßenswerte Angelegenheit. In der Einleitung wird eine kleine Einstimmung zum Gegenstand und zur Entwicklung der Paläontologie gegeben Auf 440 Seiten sind etwa 6500 Begriffe dieser Wissenschaft mit großer Sachkenntnis erläutert. Wie es im Vorwort heißt, seien allgemeine Paläontologie, Paläobotanik, Paläozoologie der Wirbellosen und Wirbeltiere und Mikropaläontologie „gleichmäßig berücksichtigt". Das ist aber bei näherem Hinsehen nicht der Fall Das Übergewicht der Palaozoologie, die bisweilen mit der Paläontologie gleichgesetzt wird, ist deutlich. Die Paläobotanik kommt eindeutig zu kurz und wird zu sehr auf das Palaophytikum ausgerichtet. Von 112 Abbildungen sind nur 13 paläobotanischen Inhalts, und auch diese sind nicht gerade glücklich ausgewählt Wichtige Begriffe der Paläobotanik des Känophytikums fehlen, das gilt sowohl für die Tertiär- als auch für die Quartärpaläobotanik. Diese Tatsachen zeigen, daß ein einzelner Autor heute schon nicht mehr in der Lage ist, bei einer sich so schnell entwickelnden Wissenschaft den Überblick zu behalten. Dennoch, für den Benutzerkreis, an den sich das Buch wendet (Studenten der Geologie, Palaontologie, Biologie, Wissenschaftler der Geo- und Biowissenschaften, Liebhaber-Paläontologen und Laiensammler), ist das Wörterbuch ein großer Gewinn. Ausländische Interessenten mit geringeren Kenntnissen in der deutschen Sprache wird es ebenfalls dienlich sein. Ausstattung und Druck sind gut, etwas unübersichtlich die stratigraphischen Tafeln mit zu kleinen Drucklettern. Ein System der Organismen zeigt uns den heutigen Stand der Erforschung der belebten Welt und regt zum Nachdenken an.

D. H Mai, Berlin

Marianne Dörfler & Ernst Dörfler: **Zurück zur Natur?** Leipzig, Jena, Berlin: Urania-Verlag 1986. 214 S., 38 Abb., 57 Farbaufn. Ln. M 19,50

Fürwahr ein wichtiges Buch, geschrieben von einem Autorenehepaar, zwei Naturwissenschaftlern. Jeder sollte es lesen, jeden geht es an. In aller Deutlichkeit und ohne Tabus gehen die Autoren auf die vom Menschen herbeigeführten Umweltprobleme ein. Wenn nach der Lektüre der anregend verfaßten Dokumentation die Nachdenklichkeit beim Leser zurückbleibt, haben beide Verfasser das erreicht, was ihr Anliegen ist Sie wollen uns vor Augen führen, wohin fehlendes Umweltdenken, ein Raubbau an der Natur und die Mißachtung der Naturgesetze in wenigen Jahrzehnten fuhrte und weiter führen wird, wenn sich nicht auf vielen Gebieten des Lebens das ökologische Denken durchsetzt. Manche der Daten und Fakten erschrecken, sei es die Zusammensetzung des Mülls, das Aussterben von Tier- und Pflanzenarten, der auf Verbrauch orientierte Lebensstil, der Teufelskreis um die Pflanzenschutzmittel oder das Ausmaß der Ausbeutung der nicht regenerierbaren Rohstoff- und Energieressourcen der Erde, um nur einige Beispiele zu nennen Anschauliche Graphiken, Schemata und überzeugende Farbfotos erganzen den Text, der sich vornehmlich mit dem Wald, dem Boden, der Luft und dem Wasser bzw mit dem weiten Feld des Umweltschutzes beschaftigt (lediglich das Lärmproblem wurde ausgelassen).

Es gibt viel zu tun, um die Fehler dieses Jahrhunderts, wenn überhaupt moglich, wieder gutzumachen Ein Zurück zur Natur ist Utopie, nicht aber ein auf Vernunft begründetes Verhältnis zu ihr. Wünschenswert ist fur die nächste Auflage ein Quellenverzeichnis.

G. Breitig, Berlin

Klaus Hanel **Biologische Abwasserreinigung mit Belebtschlamm.** Reihe „Umweltforschung". Jena: VEB Gustav Fischer Verlag 1986. 293 S, 71 Abb. auf 16 Tafeln, 46 Tab. Brosch. DDR M 47,70/Ausland DM 57,—

Der Autor, seit Jahren in der Forschung und Praxis der biologischen Abwasserreinigung tätig, stellt die naturwissenschaftlichen Grundlagen des Belebtschlammverfahrens in den Mittelpunkt seiner Publikation. Während Abhandlungen zur Abwasserreinigung zumeist die technologische Seite hervorheben, behandelt diese Schrift in hervorragender Weise in 13 Kapiteln die beim Belebtschlammverfahren ablaufenden biochemischen und biophysikalischen Prozesse, die auftretenden Organismengesellschaften und ihre Rolle beim Abbau der Abwasserstoffe, die Bewertung der Abwasser- und Belebtschlammeigenschaften, den Faktor Sauerstoff, die Effekte der Nachbehandlungen, wie z. B. die Klärung des gereinigten Abwassers, dessen Desinfektion, die Nährstoffelіminierung und die Überschußschlammbehandlung, das Betreiben der Anlagen und die notwendigen Maßnahmen bei Betriebsstorungen sowie weitere naturwissenschaftliche Aspekte.

Der Autor darf sicher sein, daß er mit der Schrift das Bestreben der Wasserwirtschaft zur Optimierung bestehender und geplanter Abwasserreinigungsanlagen sehr unterstützt und den Technikern, Biologen und Projektanten, aber auch den Studenten, vielseitige theoretische Kenntnisse für die Umsetzung in der Praxis vermittelt. Dem Verlag ist zu danken, daß er diese Publikation in die Reihe „Umweltforschung" aufgenommen hat und sich damit abermals für den Gewässerschutz eingesetzt hat.

G. Breitig, Berlin

ISBN 3-05-500429-9
ISSN 0075-9511

Erschienen im Akademie-Verlag Berlin, DDR-1086 Berlin, Leipziger Straße 3—4

Lizenznummer: 202 · 100/492/88
Printed in the German Democratic Republic
Gesamtherstellung VEB Druckerei „Thomas Muntzer", 5820 Bad Langensalza
Umschlaggestaltung. Wolfgang Janisch
LSV 1305
Bestellnummer· 763 777 0 (2111/19/1)
02400

Inhalt/Contents

Band 19 der LIMNOLOGICA umfaßt zwei Hefte (Volume 19 of LIMNOLOGICA contains two numbers).
Band 19, Heft 2, wird folgende Arbeiten enthalten (Volume 19, number 2, will contain the following contributions):
FLÖSSNER, D.: RICHARD KOLKWITZ (1873—1956) und das Saprobiensystem — RAIDT, H., & R. KOSCHEL: Morphology of Calcite Crystals in Hardwater Lakes. — CASPER, S. J., W. SCHEFFLER, K. AUGSTEN & TH. PESCHKE: Some Observations on the *Stephanodiscus hantzschii*-group (Bacillariophyta) in Waters of the GDR. III. *Stephanodiscus minutulus* or *S. parvus* in Lakes Dagow, Stechlin, Breiter Luzin, Nehmitz, Tollense and Haussee? — UHERKOVICH, G.: Weitere Beiträge zur Kenntnis der Algenvegetation auf der Sedimentoberfläche im Balaton (Plattensee, Ungarn). — KRIENITZ, L.: Algologische Beobachtungen in Gewässern des Biosphärenreservates „Steckby-Lödderitzer Forst" (DDR). — GERSTÄDT, G.: Beitrag zur Benthosflora der Recknitz im Norden der Mecklenburger Seenplatte (DDR). — VÖGE, M.: Tauchuntersuchungen der submersen Vegetation in skandinavischen Seen unter besonderer Berücksichtigung der Isoetiden-Vegetation. — KŁOSOWSKI, S.: Ökologische Amplitude und Zeigerwert der häufigeren Röhrichtgesellschaften im nordöstlichen Teil Polens — SAHA, L. C., & B. PANDIT: Variation in the Macrophytic Biomass between a Lentic and a Lotic System in Relation to Abiotic Factors. — BABENZIEN, H.-D.: Bakteriologische Parameter und Biomonitoring von Gewässern: Ein Standortvergleich. — CASPER, P., CH. BABENZIEN & G. PROFT: Untersuchungen zum mikrobiellen Abbau von pflanzlichem Detritus in Seen. — KASPRZAK, P., R. KOSCHEL, U. STEINER & K. METZDORF: "Enclosure" Experiments in Food-web Manipulation. — MODENUTTI, B. E., & M. C. CLAPS: Monogononta Rotifers from Plankton and Periphyton of Pampasic Lotic Environments (Argentinia). — PROFT, G., & W. RAUSCH: Probengröße — Partikelgröße — Fehlergröße: Zur Analyse partikulär gebundener Wasserinhaltsstoffe. — Rezensionen.

ISBN 3-05-500429-9
ISSN 0075-9511

Bei Fragen zur Produktsicherheit wenden Sie sich bitte an:
If you have any questions regarding product safety,
please contact:

Walter de Gruyter GmbH
Genthiner Straße 13
10785 Berlin
productsafety@degruyterbrill.com